ERASMUS DARWIN

Frontispiece Joseph Wright of Derby, *Portrait of Erasmus Darwin* (c. 1792).
© Derby Museum and Art Gallery.

ERASMUS DARWIN

Sex, Science, and Serendipity

PATRICIA FARA

OXFORD
UNIVERSITY PRESS

OXFORD

UNIVERSITY PRESS

Great Clarendon Street, Oxford, OX2 6DP,
United Kingdom

Oxford University Press is a department of the University of Oxford.
It furthers the University's objective of excellence in research, scholarship,
and education by publishing worldwide. Oxford is a registered trade mark of
Oxford University Press in the UK and in certain other countries

© Patricia Fara 2012

The moral rights of the author have been asserted

First published 2012

Impression: 1

British Library Cataloguing in Publication Data
Data available

Library of Congress Cataloging in Publication Data
Data available

ISBN 978-0-19-958266-2

Printed in Great Britain by
Clays Ltd, St Ives plc

For my Serendipitous Poet

ACKNOWLEDGEMENTS

ACKNOWLEDGEMENTS

Many people have helped me in the course of this project, but I should particularly like to thank James Delbourgo, Roger Gaskell, Stuart Harris, Desmond King-Hele, Nigel Leask, Tim Lewens, Martin Priestman, and an anonymous reviewer for their advice and contributions. I have also received invaluable advice and support from Tracy Bohan, my agent at Andrew Wylie, from Matthew Cotton and Emma Barber at Oxford University Press, and from my meticulous proofreader and copy-editor, Andrew Hawkey and Richard Mason. Above all, I am extremely grateful to Clive Wilmer for reading a draft version so thoroughly. The mistakes, opinions and fictionalized versions of research procedures are my own.

CONTENTS

CONTENTS

LIST OF ILLUSTRATIONS

INTRODUCTION: SERENDIPITY

It is easy to decipher *now* a past destiny; but a destiny in the making is, believe me, not one of those honest mystery stories where all you have to do is keep an eye on the clues.

Vladimir Nabokov, *Lolita*, 1955

Lolita is not the obvious starting point for a book on progress and poetry in the eighteenth century. To be honest, there was no precise beginning, no definitive moment when I decided to focus on Erasmus Darwin. I can, however, pinpoint exactly when I resolved to jettison convention and experiment with other ways of writing history. It happened unexpectedly on a late train back from London to Cambridge, when I was drowsily reading Vladimir Nabokov's novel about the experiences of Humbert Humbert with his landlady's precocious daughter. Suddenly I jolted awake.

I went over these words again. 'It is easy', I had just read, 'to decipher *now* a past destiny; but a destiny in the making is, believe me, not one of those honest mystery stories where all you have to do is keep an eye on the clues.'[1] That, I thought, is what historians actually do—as compared with what they appear to do. In their books, they tell a story that proceeds logically towards its conclusion as if they had worked out the entire plot in advance. In reality, they set off with only a dim perception of their destination, and during the course of their research they stumble between discoveries and insights, between misunderstandings and forgetfulness, between lucky finds and blind alleys. Only retrospectively do they marshal past events into a logical order.

Back in the mid-seventeenth century, during the early decades of the Royal Society, experimenters wrote in the first person, encouraging their readers to enter the laboratory and witness proceedings for themselves, even if only in their imaginations. When Isaac Newton described his optical research, he started by evoking the atmosphere of his surroundings. Working in a darkened room, he watched a beam of sunlight pass through a small hole in his window shutter to fall on a prism and cast a rainbow pattern on the opposite wall. He included himself in his account: 'I held a white Paper...And I found that...'. Such first-hand reporting of personal experiences gradually went out of fashion. Instead, scientists have adopted a dry, terse style, and exclude themselves from the scene of operations. A modern Newton would write something more like 'A coloured spectrum was produced by a light source. It was shown that...'. By using the third person and the passive voice in this way, scientists conceal their existence as variable, temperamental individuals, instead presenting themselves as objective instruments able to record results that are universally and eternally true.

Historians of science—and here I plead guilty by including myself—also like to lend themselves authority by delivering abstract generalities as if from some higher plane of expertise. This gives rise to a troubling contradiction. On the one hand, we dedicate our professional lives to undermining the notion that scientists soar above ordinary concerns as they ruthlessly and dispassionately decipher nature's secrets. We seek to demolish that convenient mythology of superiority by exposing how science is, like any other human activity, riddled with subjectivity, ambition and political intrigue. Yet at the same time, we tell our stories in the third person as though we were ourselves detached, neutral observers. By adopting this scientific rhetoric, we replicate the behaviour we are challenging.[2]

Sitting on that night train back to Cambridge, *Lolita* temporarily forgotten, I determined to abandon any such pretence of omniscience. I would conduct my readers with me on a journey through my research, and share with them—with you—the excitement of making unexpected finds, the satisfaction of tracking down those final snippets of evidence that confirm a hunch. This would be a book that would be not only about Erasmus Darwin, but also about the trials and pleasures of research. I determined that I would reveal the foggy confusion in which historians work (I assume I'm not alone in this) as they try to make sense of what they uncover. As Nabokov put it, destinies are made, not deciphered. When I embarked on narrating my Darwinian detective story, I too would be ignorant of its denouement.

Even so, this book comes with a self-protective disclaimer. Although I hope that all my historical facts are right, I have fictionalized my own activities and in places exaggerated my ignorance, naivety and incompetence. When Humbert Humbert published his pseudo-autobiography, nobody assumed that Nabokov had written a literal version of his own involvement either with nymphettes or with the United States of America. Similarly, in this book about Darwin, my personal narratives are not necessarily completely true: I am not quite such an innocent anthropologist as I make myself out to be.

And that proviso covers these introductory paragraphs.

Serendipity

Appropriately enough, the very word 'serendipity' entered the English language by a circuitous, accidental route, one that I came across by chance when reading about a book that had itself mysteriously languished unpublished for sixty years.[3]

In the middle of the eighteenth century, Horace Walpole—prime minister's son, gothic novelist, malicious gossip—happened upon an anonymous fairy tale about the three princes of Sarendip/Serendib, the ancient names for Sri Lanka. Their extraordinary powers of deduction surpassed even those of Sherlock Holmes: their feats included correctly inferring that an unseen camel was blind, lame, ridden by a pregnant woman, and missing a tooth. Freely adapting the original, Walpole concocted a salacious account of an aristocratic sexual intrigue being deviously unmasked at a dinner party—cue his invention of 'serendipity', discovering something by chance that you are not specifically looking for.

Two and a half centuries went by before serendipity was taken seriously, even though the idea was already there, waiting to acquire its name. Its most famous exponent was the French chemist Louis Pasteur, who taught his students that 'Chance favours the prepared mind'. By this maxim he meant that if you study normality carefully, you're more likely to pick up those unusual clues—the dog who fails to bark in the night, the camel who eats the grass on only one side of the road—that can help you recognize the significance of the unexpected. The classic scientific example is Alexander Fleming's refusal to throw away a Petri dish affected by an unknown mould, the penicillin that had flown in through the window. There are also some striking literary examples. For example, in the 1950s an expert on T. S. Eliot's poetry rented a house in Provence, where he happened to pick up an obscure ghostwritten autobiography by an Austrian aristocrat. To his astonishment, he realized that he had stumbled across the source for some of Eliot's lines near the beginning of *The Waste Land*: Marie, the archduke's cousin, really had existed, and phrases from her memoir of minor court scandals had lodged themselves in Eliot's mind.[4]

Serendipity launched me on this book about Erasmus Darwin. A few years ago, I was investigating the notorious 1860 debate at Oxford on evolution between Thomas Huxley (known as 'Charles Darwin's bulldog' because he campaigned so vociferously on Darwin's behalf) and his eloquent opponent Bishop Samuel Wilberforce (nicknamed 'Soapy Sam'). Allegedly, having failed to defeat Huxley on scientific grounds, Wilberforce resorted to enquiring whether it was from his grandfather or his grandmother that he was descended from a monkey. Wilberforce always maintained (wrongly) that this sally had won the argument for him, but thirteen years later Huxley got his final revenge: after Wilberforce fell off his horse and died, Huxley supposedly quipped that his brains had at last met reality, with fatal results. Although these apocryphal stories had been told many times, I hoped to find a new angle, a new way of thinking about evolution's history, by returning to the original documents and looking at them for myself.

Seeking an air-conditioned refuge during an exceptional British heatwave, I retreated to the University's Rare Books Room and started reading Wilberforce's delightfully vituperative review of Darwin's *On the Origin of Species* (1859). Like many Victorian preachers, Wilberforce wrote in a leisurely style—after all, why restrict yourself to one sentence when you can pontificate for three? Conscientiously persevering to the end of his hatchet job, I came across a slightly different version of his jibe at Huxley, in which he sniped at Darwin's ancestry by accusing him of having inherited ridiculous opinions from his 'ingenious grandsire', Erasmus

Darwin. To reinforce his derision, Wilberforce reproduced a long extract from a work published in 1798, over sixty years earlier—a poem called 'The Loves of the Triangles'. It would, he pronounced, be familiar to many of his readers.[5]

It certainly wasn't familiar to me. I migrated to the computer room and only a few minutes later had come across almost forty newspaper references. Since I found those so quickly, how many more must there be in other databases and in periodicals that have not yet been digitized?[6] What surprised me was not my ignorance of the poem, but that long after it appeared, it was being acclaimed by scientists and men of letters as mockery 'of the highest degree of merit', a prime example of 'true wit and humour [whose] fun can never be exhausted'.[7] Like Wilberforce, these writers clearly assumed their audiences would immediately know what they were talking about.

Since the vast majority of my readers will never have encountered 'The Loves of the Triangles', here are a few typical lines minus their footnotes (which you can find in the Appendix, where I've reproduced the entire poem):[8]

> And first, the fair PARABOLA behold,
> Her timid arms with virgin blush unfold!
> Though, on one *focus* fix'd, her eyes betray
> A heart that glows with love's resistless sway... (110)

The only straightforward clue to this parody is its title, which mocks a long poem by Erasmus Darwin called *The Loves of the Plants*. And I mean long—around 1,700 lines divided into four sections, or cantos. And that's without counting the prose footnotes. In what now seems a curious hybrid, Darwin used light-hearted and sexually suggestive verse to present the latest botanical theories of the Swedish taxonomist Carl Linnaeus. First published in 1789, *The Loves of the Plants* was so successful that it appeared repeatedly in English as well as being translated into several foreign languages. Almost overnight, Darwin was transformed from a provincial doctor into a literary celebrity. Even Walpole was enthusiastic, calling it 'the most delicious poem upon earth'.[9]

Darwin claimed to be a reluctant media star, initially publishing anonymously and telling his friends 'I court not fame, I write for money.'[10] By the time 'The Loves of the Triangles' appeared in 1798, his identity as a popular poet was well-known, and he had produced two major works in verse, *The Loves of the Plants* and *The Economy of Vegetation*, sold bound together into a single volume called *The Botanic Garden*. Just as important was *Zoonomia* (1794–6), a hefty two-volume medical textbook highly praised for its clinical information and insights, but also containing Darwin's controversial suggestion that life and the cosmos have evolved over time.

According to a close friend, Darwin admired 'the wit, ingenuity, and poetic merit' of 'The Loves of the Triangles'.[11] However, this may well have been a brave front. The *Anti-Jacobin* issued its satire in three instalments over a period of almost a month. As the weeks went by, the humour became darker and the indictments more savage. For the next four years, Darwin continued to write, but after this public mauling, he threw away entire passages of his next long poem and substantially rewrote it. Eventually, the final manuscript was published a year after his death under a revised title, *The Temple of Nature*.[12]

A hundred years later, Darwin's originals had been forgotten, whereas 'The Loves of the Triangles' was still being praised as 'one of the very finest examples of English burlesque'.[13] Scanning through the digitized pages of Victorian periodicals, I learnt that it had been a political satire orchestrated by supporters of William Pitt, the Tory prime minister. Committed to protecting the British establishment from Napoleon Bonaparte and the French Revolutionaries, his protégés had produced a short-lived journal called the *Anti-Jacobin; or, Weekly Examiner*, which featured regular satirical poems as well as some savage caricatures by James Gillray.

This further information prompted further questions. What did their conservative Francophobia have to do with Erasmus Darwin, whom I knew as a paternalistic Midlands doctor with a passion for botany and poetry? Seventy years later, why did *Punch* refer to 'the moral character of a parallelogram'?[14] And why, I wondered, did several Victorian writers choose to reproduce this particular extract:

Debased, corrupted, groveling, and confined,
No DEFINITIONS touch *your* senseless mind;
To *you* no POSTULATES prefer their claim,
No ardent AXIOMS *your* dull souls inflame;
For *you* no TANGENTS touch, no ANGLES meet,
No CIRCLES join in osculation sweet![15]

Although I didn't myself find these lines inspiring, I could understand their relevance for Victorian scientists railing against abstract mathematics, or for literary critics bewailing the lack of overt emotion in a novel. But what did they have to do with Enlightenment politics? In any case, what was so witty about them?

I remembered an article by an eminent French historian, Robert Darnton, discussing his bewilderment at an eighteenth-century jape in a Parisian printing shop. For weeks, the workers had roared with laughter whenever they recounted how apprentices had bludgeoned all the local cats to death, stringing them up on an improvised gallows and dumping them down into the courtyard below. How could this ever be thought of as funny? Along the lines of Pasteur or Fleming exploiting the unanticipated event thrown up by chance, Darnton recommended exploring this alien sense of humour more deeply. Instead of dismissing such jarring jokes as despicable and incomprehensible, historians should, he advised, grasp them as opportunities to understand vanished ways of viewing the world.[16]

I decided to follow Darnton's advice and investigate more fully this poem that seemed so painfully unfunny to me. My first move was to ask some experts—the academic equivalent of 'phone a friend'. A poet with a particular interest in Victorian literature had never heard of 'The Loves of the Triangles', but was gratifyingly—if somewhat unconvincingly—enthusiastic about the possibilities for rewriting the accepted canon of English comic verse in the nineteenth century. (This poet may well regret his initial interest, as he subsequently became a major source of information.) Similarly, a specialist in Victorian evolution suggested plenty of books to read, but had no idea that Wilberforce had supplemented his smear tactics with Enlightenment poetry.

By scouring the indexes of all the books I could find on Erasmus Darwin, I soon established that 'The Loves of the Triangles' had dramatically affected Darwin's reputation, converting him from a respected poet into a national laughing-stock. Despite the importance of that impact, none of his biographers had written more than a few sentences about the *Anti-Jacobin* parody.

A few weeks later, serendipity came my way again. Browsing in Amazon, I found that I could buy an 1807 edition of the collected *Anti-Jacobin* poems—including 'The Loves of the Triangles'—for under £10. This book was evidently no rarity. Mine turned out to be the fifth edition in eight years, which indicates how popular they were at the time—and its battered condition suggested frequent reading. An unanticipated demonstration of my poem's wide influence came when the *Oxford English Dictionary* (henceforth *OED*, an indispensable online research tool) told me that 'osculating' originally meant kissing, but later became a geometrical term for touching—and the earliest known example is from 'The Loves of the Triangles'.

To my delight, when my Amazon copy arrived, it included a handwritten note telling me that 'These poetical effusions [were] republished in a splendid quarto volume, the expense of printing which was defrayed by Mr Pitt, Mr Canning, nearly all the Cabinet Ministers, & many others connected with that party'. The antics of ardent Axioms and debased Definitions clearly carried great political importance. But why? As I investigated this poem's history, and tried to grasp the significance of Latin jokes and contemporary gossip that would then have been immediately self-evident, I gradually came to appreciate its witticisms, although many are probably now too distant and too complex to unravel fully.

Four Poems

One of the most difficult aspects of writing a book is devising its structure. Somehow, you have to present a mass of tangled information in a straight line without repeating yourself. The great temptation is to put everything into the introduction, but then you find that what should provide a brief taster has expanded into the length of a book. To make matters more difficult, publishers will only accept proposals that have been organized into neat, self-contained chapters. How do you do that before carrying out the research?

All writers are familiar with the logical impasse I faced. How could I discuss this satire on Darwin without first describing his own poetry? But without presenting the satire, how could I explain why I was choosing particular extracts from Darwin to examine? To resolve this dilemma, I decided to organize my book around four poems: one by Darwin's parodists, and three far longer ones by him. As the Contents page shows, for each one I identified a general theme particularly relevant to that poem. The first section focuses on 'The Loves of the Triangles'—but for readers new to his life, work and poetry, I also summarize Darwin's achievements and introduce some contemporaries. Then I move chronologically through Darwin's three major poems, focusing for each on one of the Big Questions: Sex, Politics and Race (the other Big One, Religion, also pervades this study).

While thinking about 'The Loves of the Triangles', I travelled on a serendipitous journey through the eighteenth century that took me to some unexpected places, but as a bonus, revealed a great deal about people's lives, prejudices and experiences. Guided by the discoveries I made along the way, my project grew, repeatedly shifted direction, and ended up being very different from a conventional biography. Although not a systematic study of Darwin's poems, my exploration is more thorough than any previous one. And although I'm a historian of science, there's nothing technical in this book—definitely no equations or diagrams.

This book presents my conclusions, but it also tells a story about how I reached them, because I hope to give readers an impression of how research actually happens. Academics are neither omniscient nor infallible, and my literary expedition includes many backtracks, missed signposts and circuitous detours. If you do reach its final destination, I hope you'll agree that although it was his grandson who became world-famous, Erasmus Darwin also played a crucial role in British science, politics and literature.

THE LOVES
OF THE TRIANGLES

POETRY, GEOMETRY,
AND SATIRE

The most unfailing herald, companion, and follower of the wakening of a
great people to work a beneficial change in opinion or institution, is
poetry.... Poets are the hierophants of an unapprehended inspiration; the
mirrors of the gigantic shadows which futurity casts upon the present; the
words which express what they understand not; the trumpets which sing to
battle, and feel not what they inspire; the influence which is moved not, but
moves. Poets are the unacknowledged legislators of the world.

Percy Bysshe Shelley, 'In Defence of Poetry' (1821)

Erasmus Darwin could be a dangerous man to know. In February
1793 the King of France had just been guillotined, France had
declared war on Britain, and the government was clamping down
on radical campaigners advocating political reform. 'In the present ferment of men's minds,' wrote Darwin's admirer William Stevens, headmaster of a nearby school, 'were I to visit him frequently we should be supposed
to be plotting sedition.'[1]

This was not a view of Darwin that I recognized. I came across it while
I was enjoying the deadpan humour of Stevens's diary, which is packed
with informative if acidulous comments on his Derbyshire neighbours
(Mrs Greaves was summed up for posterity as an eccentric wife whose
'Constitution is as cold as a Greyhound's nose').[2] Darwin flits in and out
of these pages under some stereotypical guises—the rebarbative dinner
guest, the diplomatic doctor, the arrogant author, the obsequious suitor,
the ingenious inventor, the enthusiastic botanist. But Darwin the political
agitator? Him I had never encountered.

Darwin is usually portrayed as a representative of comfortable, provincial stability. A couple of years before 'The Loves of the Triangles' appeared,
Samuel Taylor Coleridge made a special detour to visit the elderly doctor
at his home in Derby. For the young would-be poet, Darwin was one of

the city's main attractions, who 'possesses, perhaps, a greater range of knowledge than any other man in Europe, and is the most inventive of philosophical men'.[3] Energetic and sociable, this corpulent teetotaller had run a successful medical practice in Lichfield, was a Fellow of London's Royal Society, and acted as a focus for the Lunar Society, an informal Midlands network of factory owners, social reformers and scientific colleagues. The father of twelve children by three women (two wives and his son's governess), Darwin somehow found the time to be a prolific writer: his publications included textbooks on medicine, botany and agriculture as well as three tomes of poetry.

Why should a headmaster be so cautious about visiting such a reputable man? Darwin clearly had another side—or perhaps several. He was, I discovered, a co-founder of the Derby Society for Political Information, a local organization that hit the London press after its representatives presented a manifesto to the French National Assembly demanding the vote for all adult males. The editors of the *Morning Chronicle* were prosecuted for printing a copy of this inflammatory document, which was 'rumoured to come from the pen of a writer, whose productions justly entitle him to rank as the first poet of the age;—who has enlarged the circle of the pleasures of taste, and embellished with new flowers the regions of fancy'—in other words, Erasmus Darwin.[4]

Darwin escaped lightly. In the years following the French Revolution, the British judiciary clamped down to ensure that sedition did not seep across the Channel, and some victims of the repressive regime had already been deported to Botany Bay. In this trial, despite the judge's unprofessionally heavy hints pushing the jury towards culpability, after fifteen hours the editors were released. Without that 'Not Guilty' verdict, perhaps Darwin would have been shipped over to Australia, and perhaps he would not have had a grandson, and perhaps the history of evolution would have been different, and...

'What if...' questions are tempting but risky and not necessarily fruitful. After all, what happened is what happened. On the other hand, there are always new decisions for historians to make about which events to focus on, what sort of pattern to weave, and—most difficult of all—what to leave out. Researching into the past entails not only accumulating knowledge, but also divesting yourself of misleading preconceptions. The people I'm trying to understand and write about had no idea what the future held in store, whereas my perception of them is biased by knowing how important science and industry became during the next couple of hundred years. Uncovering the roots of that development is important—but focusing exclusively on those aspects of their activities is to distort the past. Although Darwin as radical activist was news to me, perhaps to him and his colleagues this role meant more than the botanical poetry, industrial innovations and ingenious inventions for which he is now celebrated.

Speaking on behalf of detectives everywhere, the novelist G. K. Chesterton stressed that it's not hard finding the solution: the real problem is working out the right question. For a start, I asked myself why the *Anti-Jacobin* poets had picked on Darwin as an important satirical target. My first steps seemed clear: I needed to find out more about Darwin, and to study 'The Loves of the Triangles' more systematically. The next two chapters describe my initial discoveries.

CHAPTER ONE

Erasmus Darwin

As I was turning over some old mouldy volumes, that were laid upon a Shelf in a Closet of my Bed-chamber; one I found, after blowing the Dust from it with a Pair of Bellows, to be a Receipt Book, formerly, no doubt, belonging to some good old Lady of the Family. The Title Page (so much of it as the Rats had left) told us it was 'a Bouk off verry monny muckle vallyed Receipts bouth in Kookery and Physicks.' Upon one Page was 'To make Pye-Crust,' and in another 'To make Wall-Crust,'—'To make Tarts,'—and at length 'To make Love.' 'This Receipt,' says I, 'must be curious, I'll send it to Miss Howard next Post, let the way of making it be what it will.'

Letter from Erasmus Darwin to his future wife, Mary Howard
(24 December 1757)

When I visited Lichfield, the Midlands city where Darwin practised as a doctor for quarter of a century, I discovered that the National Heritage Industry has missed a marketing opportunity. For a place with such a rich history and magnificent buildings, Lichfield is surprisingly unappealing to tourists. The once-splendid coaching inn has been taken over by a down-market American chain, and the smoke-darkened bricks confirm that Lichfield sits near the bottom of the country's ecological league tables. Although now overshadowed by Birmingham, fifteen miles away, Lichfield was a thriving commercial and cultural centre in Darwin's time. Other distinguished residents included Samuel Johnson, who boasted that 'we are a city of philosophers; we work with our heads, and make the boobies of Birmingham work for us with their hands'.[5]

Not everything has deteriorated. The city is still dominated by the massive medieval cathedral, which has some unusually fine stained glass and is unique in Britain for its three spires. Severely damaged during the Civil War, it was restored to its present splendour during the Victorian era. Darwin knew the building in a semi-ruined state, its intricate stonework plastered over with cement, and he helped to prevent further deterioration by installing one of England's earliest lightning conductors.

Darwin's house backs on to the Close, a higgledy-piggledy cluster of medieval brick cottages dwarfed by the cathedral's sandstone walls. Although his comfortable family home has been converted into a museum, its main features have been preserved, so you gain some impression of what it would have felt like to live there. When you go into the lovingly restored garden, you are immediately aware of the cathedral's looming Gothic presence. At night, Darwin was struck by 'the shadow of the three spires... the moon rising behind it, apparently broken off, and lying distinctly over our heads as if horizontally on the surface of the mist, which arose about as high as the roof...'[6]

As well as this physical prominence, cathedral affairs were central to Darwin's emotional life. His major patron in Lichfield was a canon, and his father-in-law was a church official living in the Close. Several of Darwin's children were christened in the cathedral, while the babies who died were buried in its graveyard. Darwin himself asked to be placed in a church near Derby next to his second son, who had been found drowned in a river at the age of forty (almost certainly suicide, tactfully covered up). Even so, the congregation of Lichfield commemorated Darwin as one of their own, installing in the cathedral aisle a handsome marble plaque and brass plate paying tribute to their much-loved local doctor, inventor, and poet.

Agreeing on the inscription's precise wording must have involved some diplomatic discussions. Encapsulating somebody's life in a few sentences is always tricky, but how much more so in a cathedral. Darwin held unorthodox theological views and became notorious for challenging the Bible's account of creation; moreover, he fathered two illegitimate children, wrote flirtatious poetry, and regarded sexual energy as the driving force of the universe. Such inconvenient details had been discreetly excluded from the inscription's glowing account of an upright citizen. How, I wondered, would I write about Darwin myself? The night before, I happened to have been reading a novel that opens by challenging the standard approach to biography: 'If you really want to hear about it, the first thing you'll probably want to know is where I was born, and what my lousy childhood was like, and how my parents were occupied and all before they had me, and all that David Copperfield kind of crap, but I don't feel like going into it, if you want to know the truth.'[7] The more I thought about Darwin, the more I realized I didn't feel like going into it either.

Standing in the cathedral, contemplating his carved marble face, I thought about the drawbacks of condensing Darwin's life into a chronological sequence. Even if I could decide where to start (his birth in 1731? his parents' marriage in 1724? his family's move to the Midlands in the seventeenth century?), narrating events in the order they happened to him seemed neither helpful nor interesting. Like many people who end up being famous, on a day-to-day basis Darwin led a muddled life that, at the time, didn't seem to be heading in any particular direction: only in retrospect can anyone lay claim to being born with the inevitable destiny of discovering X-rays or being elected president of the United States. However much Darwin fascinated me, I had no desire to sink myself into the minutiae of his day-to-day existence.

Deciding what to leave out can reveal as much about biographers as their subjects. How could I presume to describe Darwin when not even those who knew him could agree? It seems that people 'either exalt him as having been almost superior to human frailty, and exclaim with Sir Brooke Boothby—"Darwin was one of the best men I have ever known;" or stigmatise him as an empiric in medicine, a Jacobin in politics; deceitful to those who trusted him, covetous of gain, and an alien to his God.'[8] Rather like estate agents' descriptions, apparent compliments can be double-edged. According to a lifelong friend, Darwin 'despised the monkish abstinences and the hypocritical pretensions which so often impose on the world'.[9] Since he was grossly overweight even by eighteenth-century standards, should I interpret that as meaning 'Darwin enjoyed over-indulging and had no qualms about speaking his mind frankly'?

On the other hand, I needed a framework for slotting his achievements into the panorama of his existence. Darwin once recalled wandering down the cathedral aisle in Lichfield and examining the sculpture of a recumbent man whose feet and head are visible in two adjacent niches. Even though a solid stone wall fills the space where his body should have been, Darwin's imagination bridged the gap to show him the entire figure. Similarly, he wrote, poets and painters evoke powerful images by showing only a part of the whole.[10]

On that first visit to Lichfield, I too held only limited views of Darwin. I had read about his activities as a doctor who was proud of his provincial heritage and participated in local scientific networks. And I knew that later in his life, he turned to writing books about science, three of them in the form of poems. Resembling the statue's artificially separated feet and head, these two disjointed perspectives—the physician and the author—were the only glimpses I had. These are what I describe here—and in the rest of this book I venture behind that wall to paint fuller pictures of what had previously lain between them, concealed from my gaze.

Erasmus Darwin as Provincial Physician

First, the bare details of his origins. Erasmus Darwin was born on 12 December 1731 at Elston Hall, near Nottingham. His father Robert was a lawyer, and Erasmus was the youngest of seven children, all of whom survived. His mother, Elizabeth Hill, lived until she was ninety-four: that might seem only meagre information, but unlike many wives and mothers, at least her name and existence are recorded.

Next, some notable aspects of Darwin's childhood. An attack of smallpox left him with a badly scarred face, a servant hit his head so hard that a lock of hair became permanently white, and he stammered badly, a characteristic he passed on to his son and famous grandson. While his three sisters stayed at home, Darwin and his three brothers were sent to the best nearby boarding school, and then to St John's College, Cambridge. Already interested in poetry ('My dearest Sue/Of lovely hue/No sugar can be sweeter...'), Darwin completed his training at Edinburgh, Britain's leading medical school.[11]

Darwin's first attempt to earn his own living was a disaster. In 1756, the summer he qualified as a doctor, he set up practice in Nottingham without realizing how important it was to find a patron who would recommend him to prospective patients. After only three months, Darwin migrated to Lichfield, and this time used his contacts to set up introductions in advance. Soon he established a wide social circle, although he never got on well with Johnson, also a large domineering man who liked to be the centre of attention. But he did become very friendly with the family of Thomas Seward, a canon at the cathedral who organized literary evenings at his home in the Bishop's Palace.

Seward's daughter Anna was only thirteen years old when Darwin first met her, but she already loved writing poetry and later become renowned as the literary 'Swan of Lichfield', the 'Queen Muse of Britain'.[12] For many years they maintained a close if stormy relationship, coloured by strong sexual tensions and soured by her accusations that he had stolen her verses. Bitterness makes her an unreliable source, although in remembering him as a blundering but charming bully, her judgement fits well with that of others. 'Florid health, and the earnest of good humour, a sunny smile, on entering a room, and on first accosting his friends, rendered, in his youth, that exterior agreeable, to which beauty and symmetry had not been propitious,' she started generously, before twisting the knife to focus on his stammer and the condescending manners of a man who believed himself to be superior: 'he became, early in life, sore upon opposition...and always revenged it by sarcasm of very keen edge'.[13]

Only a few weeks after his arrival in Lichfield, Darwin struck lucky. By restoring to health a rich young man pronounced sick beyond hope, his reputation as a miraculous healer started drawing in the patients he needed. Within four years, his life had changed dramatically. He was now the husband of Anna Seward's friend Mary Howard (known as Polly), the father of their two sons (the engagement was suspiciously brief), and the owner of a handsome house, solidly symmetrical in the Georgian style. Steadily building up his practice, Darwin settled down to the contented life of a prosperous married physician.

Darwin's predecessor in Lichfield had been a local historian with an extremely rich wife but no medical degree. In contrast, intensively trained at Edinburgh's progressive Medical School, Darwin belonged to a younger generation, and he soon gained the reputation of being an outstanding doctor. His patients included Georgiana, Duchess of Devonshire, and King George III apparently invited him (unsuccessfully) to London several times, hoping that Darwin might cure the royal bouts of insanity. Even allowing for some exaggeration, Darwin genuinely seems to have preferred working provincially to becoming a fashionable doctor in the metropolis. Similarly, his friend Joseph Wright became nationally famous as an artist, but spurned London to remain in his native Derby. Thanks to men like these, as labourers moved towards the new Midlands factories, Britain's cultural life expanded beyond London and the south.

Well-provided with servants, Darwin had sufficient spare time to carry out experiments, devise inventions, and sketch out plans for local improvements. Centrally placed, he participated in two scientific networks. The more prestigious was London's Royal Society, where he was elected a Fellow in 1761. By then, the Society's international journal, the *Philosophical Transactions*, had already carried his accounts of original research into gases and lung conditions; for the rest of his life Darwin continued to publish books and articles on topics as varied as squinting and farming. Although only rarely visiting London, like many provincial residents he kept up with the latest developments through letters and magazines.

In contrast to these long-distance relationships, the local but less well-known Lunar Society played an important role in Darwin's life.[14] Based in the Midlands, this informal group of friends met around once a month, often accompanied by their wives and children. Sometimes credited with launching the Industrial Revolution, Darwin's Lunar colleagues included eminent inventors and factory owners such as James Watt, Joseph Priestley, Matthew Boulton, and Josiah Wedgwood. It was with these men that Darwin discussed his inventions and his industrial plans. Worried that he might be branded an eccentric, Darwin kept quiet in public about his more unusual devices, but he did confide in his Lunar companions, exchanging long letters about his horizontal windmill and enlisting their help (unsuccessfully) to market his letter-copying machine and his oil lamp. Many proposals never reached their final stages, while quite a few— his extractor fan, his artificial bird with flapping wings, his canal-boat lift, his automatic window-opener sensitive to sunshine—remained at the sketchbook stage. Darwin was, however, a persuasive lobbyist, and his enthusiasm carried through to completion several of the canal projects that transformed Britain's landscape and made possible the country's rapid industrialization.

On a day-to-day basis, Darwin's medical career dominated his existence. He routinely travelled long distances through the countryside—several thousand miles a year, he claimed—in the two brightly painted, well-sprung carriages he had designed himself, piles of books and hampers of food wedged in alongside his medical bags. When the path was too muddy for wheels, he rode his horse, which he called Doctor. (Darwin had a great fondness for puns. This was one of his favourite jokes: 'A Miller was asked by two would-be wags who were riding one on each side of him whether he was most knave or fool. "I cannot exactly say," retorted the Miller, "but I believe I am between Both." '[15]) In addition, Darwin often prescribed and diagnosed by correspondence, and letters to his friends are packed with tips for good health—valerian for the eyes, steel filings and laudanum for rheumatism, foxglove and swinging for a teenager with tuberculosis. For just about everything else—laxatives and bleeding, the favourite Enlightenment remedies.

However fanciful his poetry, when it came to health, Darwin was an eminently sensible, pragmatic man, and many of his then-novel prescriptions are now routine: take plenty of exercise in the fresh air, abstain from alcohol, avoid too much spicy food. His excessive weight suggests that he ignored some of his own advice (his appetite for desserts was notorious), but he medicated himself so that he could get on with his life:

> ...this morning about three I waked with much pain, and tumour, and redness about the joint of the toe. I bled immediately to about 10 or 12 ounces, my pulse was not much quicken'd nor much hard. about 4 I took calomel gr. vi. [laxative] at six this operate [sic], at seven I was very faint, and had a slight chillness and 3 or 4 more stools. At 8 I put aether repeatedly on the tumid and red part, and kept it from evaporating by a piece of oil'd silk, the pain became less. At 9 set out for Burton, with difficulty got into the chaise. At 11 was easier, went on to Aston...'[16]

Darwin was a man with a conscience who took his Hippocratic oath seriously, charging according to wealth and giving free advice to those who could not afford to pay. A progressive doctor who wanted to integrate the latest scientific ideas within the conventional craft of medicine, he put great energy into assessing new treatments—smallpox vaccination, therapeutic gases, electric shocks. Even so, in the absence of antibiotics, there was often little that any physician could do except help a sufferer die comfortably, and Darwin was popular with his patients because he offered them great psychological support as well as liberally prescribing opium to numb their pain.

Despite his blustering exterior, Darwin responded sensitively to suffering and found this constant contact with disease and death hard to bear. When an outbreak of measles among the local children prevented him from attending a Lunar Society meeting, he railed at the fate of 'nine beautiful children', those 'innocent little animals' he watched 'cough their hearts up'. Such repeated reminders of human mortality made Darwin question the benevolence of a Christian God; surely, he wrote, it must be that the 'Devil has play'd me a slippery trick'.[17]

Erasmus Darwin as Scientific Author

By the summer of 1770, when Darwin was thirty-eight, this fourteen-year period of busy domesticity had come to an end. As often happens, several things went wrong at the same time. Most importantly, his wife died after a long illness exacerbated by opium and alcohol. His third son Robert—future father of Charles—was four years old, but two other babies had failed to survive, and a severe knee injury incurred during the collapse of an experimental carriage had reduced his mobility and made him even heavier.

Gradually, after enduring a period of despair, Darwin managed to regain his interest in the outside world. Involving himself in new technological projects with his Lunar friends, he dreamt up ambitious designs and also became intrigued by the botanical research of Carl Linnaeus. In 1775, after acquiring two daughters during a protracted affair with his sons' governess, he fell in love again, this time with a woman called Elizabeth Pole. She was probably introduced to him by his long-term patient Joseph Wright of Derby, who had painted portraits to decorate her large country mansion. This beautiful, energetic mother adored plants, but—unfortunately for Darwin—was already married. Nevertheless, he wooed her assiduously: he converted a nearby plot of land into a scenic garden, contrived for her to stay with him while her children were ill, and sent her passionate love poems tinged with eroticism, such as this ambiguously titled 'Hunting Song':

> When bright Eliza mounts her steed,
> With such sweet grace she rides him:
> No more their dogs the hunters heed,
> And foxes pass beside 'em....
> Ye love-struck swains, one thing I'll tell
> I fear may disconcert you:
> There is one fault attends this Belle—
> You'll hate her for her virtue.[18]

Unexpectedly, when Darwin was almost fifty, Pole's husband died. Darwin immediately intensified his courtship, and she agreed to marry him on one condition—that he leave Lichfield and move to her country house at Radburn Hall, near Derby. This decision turned out to work well for both of them, even though their lives together were marred by the death as adults of his eldest sons, Charles and Erasmus. Separated from his Lichfield and Lunar Society friends, Darwin founded the Philosophical Society of Derby, built up a new circle of scientific acquaintances, and fathered a further seven much-loved children, all of whom survived.

Ensconced in rural Derbyshire, newly married to a gardening enthusiast, Darwin turned to botany and embarked on the writing career that made him famous. Although still practising medicine, carrying out experiments, and developing inventions, he began translating Carl Linnaeus's large catalogue of plants into English from Latin, the language that was understood internationally, but only by well-educated men. Previous translators had chosen the easy option of keeping Latin names for technical plant terms, but Darwin decided to anglicize everything, and some of the labels he chose are still in use. His *System of Vegetables* (1783) is a dense scientific tract, a thousand pages long and the product of detailed investigations.

As Darwin neared the end of this self-imposed task, he started playing with the idea of using poetry to make botanical classification more accessible. Confiding in a close friend, Darwin revealed that his romantic chase after Pole had revived earlier obsessions, and the first result, in 1789, was a long poem combining gentle eroticism, mythological whimsicality, and botanical information. Reviewers raved, sales soared, and younger men—William Blake, Samuel Taylor Coleridge, Henry Fuseli, Percy Bysshe Shelley, William Wordsworth—began to absorb his influence. In contrast, many readers now find it hard to get beyond the opening lines:

> DESCEND, ye hovering Sylphs! aerial Quires,
> And sweep with little hands your silver lyres;
> With fairy footsteps print your grassy rings,
> Ye Gnomes! accordant to the tinkling strings;[19]

Publication brought national fame, but Darwin's Derbyshire life continued much as before. Even when the French Revolution erupted that July, it had little immediate effect on this provincial physician. Although there were seven children under the age of nine in the house, Darwin continued to write and research, to practise medicine and entertain his friends, to encourage his son Erasmus as a lawyer and the younger Robert as a doctor. At that time he had no idea that Erasmus would sink into morbid depression, and that Robert would have a famous son called Charles who would perpetuate his own interest in evolution. Two years later, his next major work appeared—*The Economy of Vegetation*. Half as long again as *The Loves of the Plants*, it was more ambitious in scope. Although less overtly about sex, it was written in the same florid style and intended as a companion piece on science, industrialization, and politics. Together they made up the enormously successful *Botanic Garden*.

Now sixty years old, Darwin continued writing, inventing, practising medicine—and also thinking. The commercially sensible decision would have been to set about producing *Botanic Garden Two*, but Darwin was committed to social progress. One practical contribution he could make was in medicine, by teaching people to care for themselves so that the nation's health would be improved from one generation to the next. How is it, he wondered, that living creatures stave off death and decay? How could he pass on the expertise he had gained as a successful doctor? He wanted to find out why human beings are so special, if not unique—why do they try to create beauty, to improve the world around them, to enjoy themselves?

At Edinburgh, Darwin had learnt about the research of Albrecht von Haller, a Swiss physiologist who emphasized the importance of the nervous system and studied how bodies react to stimuli. Through his experiments, Haller showed that muscle fibres contract—or as he put it, their behaviour is irritable; in contrast, nerve fibres feel—they have what he called sensibility. 'Sensibility' is one of those elusive eighteenth-century terms—'sympathy' is another—that were in widespread use but are hard to pin down. Often attributed to refined young women, sensibility embraces sensitivity, but also has an emotional component implying a powerful connection between the physical nervous system and a person's psychological state.[20]

Haller's suggestions stemmed from his scientific research, and for many people seemed to carry enormous theological threats. Traditional Christians believed in a dualistic universe, a twofold system comprising spirit and matter. In this model, inert matter—billiard balls, wood, dead flesh—is unable to move, act or think for itself. Spiritual power is completely different. Endowed by God, who is Himself pure spirit, it somehow transforms earthly substances into living beings that can feel, make decisions, and shift objects from one place to another. In contrast, materialism provided a different solution to the quandary of existence. This was the new philosophy most discussed and feared in eighteenth-century Europe. According to materialists, life is generated not by God's direct intervention, but from atoms and molecules that rearrange themselves, perhaps as the natural outcome of chemical or electrical activity. In aligning himself with Haller, Darwin went against conventional British thought.

Darwin first committed his contentious ideas to print in *Zoonomia* (1794–6), a two-volume compendium about life, physiology, and disease that summarized conclusions derived from a lifetime of practising medicine. As a textbook for treating illness, *Zoonomia* was unparalleled and highly acclaimed. Yet despite the glowing reviews, it marked a turning point in Darwin's reputation, because he also expressed publicly the thoughts about reproduction and evolution he had been pondering for years. Most British people believed that God had created the world, along with all its plants and animals, exactly as it is now. Controversially, Darwin suggested that living creatures emerged later from a single entity he called a filament, and have progressed continuously since then.

If the political situation had been different, *Zoonomia* might have attracted less attention, but when 'The Loves of the Triangles' explicitly parodied its ideas, Darwin's comfortable existence as a respectable if idiosyncratic provincial poet skidded to a halt. Critics accused him of blurring the distinction between mind and matter, of attributing life to a purely physical process rather than recognizing it as a gift from God. For him, living creatures—human beings, animals, plants—share some essential quality, some 'spirit of animation' that makes them respond to their surroundings. Unlike those Enlightenment super-rationalists who conceptualized people as living machines, Darwin celebrated the complexities and the pleasures of existence. He also sketched out tentative ideas about evolutionary change. Influenced by his medical practice, Darwin came to see life as a constant struggle for survival. His wide clinical experience convinced him that people not only transmit diseases they have inherited, but also pass on to their descendants characteristics they have acquired during their own lifetime, such as broad shoulders or alcoholism.

Two other books appeared during Darwin's lifetime. First came *A Plan for the Conduct of Female Education in Boarding Schools* (1797). This slim volume originated as advice to his two illegitimate daughters, Susanna and Mary Parker, for whom he had bought a converted inn to run as a school for girls. Supporting them so generously and openly was unusually enlightened, even though Darwin believed that they should teach their female pupils to run a household rather than a career. Along with fresh air and exercise, scientific subjects were high up on the curriculum (a handy advertising opportunity for his own books)—not so that women would be inspired to carry out scientific research, but to provide future husbands with informed companions.[21]

In contrast, *Phytologia: The Philosophy of Agriculture and Gardening* (1800) is a substantial research study (*phyto* comes from the Greek for 'plants'). Despite the strange subtitle, it is mostly an unphilosophical discussion about plants, concentrating on how to make them grow better and hence yield more profit. In principle, Darwin had chosen a marketable topic: the population was soaring, and agricultural improvements were essential for keeping people fed and able to work. Even so, this book attracted little attention. Perhaps by then he was regarded as a has-been. Or perhaps some of Darwin's schemes were too grandiose—importing sick rats from America to spread disease amongst British vermin, planting Scottish mountains with fir trees, cultivating beet to yield home-grown sugar and so relieve the slaves in Jamaica.

Similarly, *Phytologia* has held little appeal for historians. Personally, I found it pretty dreary (the title page promises information about draining morasses and an improved drill plough) until I reached the penultimate chapter, which is devoted to the splendid topic of happiness. As if fresh from a course on positive thinking, Darwin searches out the pleasures associated with sex and reproduction, even arguing for the overall benefits of death. Conscious that he was drawing near the end of his own life, watching friends and relatives of all ages die around him, Darwin consoled himself by explaining that higher animals have a greater capacity to enjoy themselves than primitive creatures. In his wishful-thinking vision of creation, evolution works by maximizing the amount of pleasure in the world. Mountains, he concludes, are ancient accumulations of once-living pleasure-seeking creatures, 'MONUMENTS OF THE PAST FELICITY OF ORGANIZED NATURE!' My other reward for persevering that far was to discover a quotation that seems to foreshadow the theories of Darwin's Victorian grandson: 'Such is the condition of organic nature! whose first law might be expressed in the words, "Eat or be eaten!" and which would seem to be one great slaughter-house, one universal scene of rapacity and injustice!'[22]

A couple of years later, in 1802, Darwin himself died, probably from a lung infection that could have been cured with modern drugs. By then, he had completed his third long poem—*The Temple of Nature*, which appeared posthumously and was the most outspoken statement of his controversial theories. In this poem, he develops still further the tentative suggestions he had made in *Zoonomia*, sketching out a progressive evolutionary model for the living world. Life, he claims, appeared in a single moment of creation many, many ages ago. Governed by natural laws in a universe charged with sexual energy, increasingly complex organisms gradually appeared, reaching their apogee in human beings.

Darwin's wife lived on for another thirty years at their final home, a renovated priory originally owned by his drowned son Erasmus. Since their grandchildren nicknamed the house 'Happiness Hall', she presumably perpetuated the compassionate exuberance that pervades Darwin's life and literature.

CHAPTER TWO

'The Loves of the Triangles'

...the growing good of the world is partly dependent on unhistoric acts; and that things are not so ill with you and me as they might have been is half owing to the number who lived faithfully a hidden life, and rest in unvisited tombs.

George Eliot, *Middlemarch* (1871–2)

'The Loves of the Triangles' appeared in the *Anti-Jacobin*, a journal I knew nothing about. The first thing I found out was that 'Jacobin' is not at all a straightforward label. 'Jacobus', Latin for the English 'James' and French 'Jacques', features in the descriptions of (at least) three important groups. First came the Jacobins, Dominican friars attached to the Church of Saint Jacques in Paris. Next were the Jacobites, British followers of the deposed King James, and nothing to do with this story. Finally, Jacobins were political activists who acquired this name because their early meetings during the French Revolution were held in the disused convent of the Dominican Jacobins. To complicate matters further, by 1798, when 'The Loves of the Triangles' appeared, 'Jacobin' had lost this specific meaning and had become a general term of abuse. In a reaction resembling anti-Communist paranoia after World War II, British people were responding to a threat that was widely perceived but only vaguely defined. Perhaps to entice affluent customers, pubs displayed signs saying 'No Jacobins Admitted Here'.[23]

In the early years of the French Revolution, many British people were sympathetic towards its goals, but their attitudes changed rapidly after the French king was guillotined in 1793 and the Reign of Terror broke out the following year. To safeguard their own positions, privileged spokesmen fed this national anxiety, encouraging Francophobia by declaring that any reform, however minor, would inevitably lead to cataclysmic disaster. Taking advantage of this fear and hostility, the eloquent Irish politician Edmund Burke warned that the comfortable British way of life risked being engulfed by a 'vast, tremendous, unformed spectre, in a far more terrifick guise than any which have ever yet overpowered the imagination, and subdued the fortitude of man'.[24] Spreading invisibly like an infectious disease, anarchy would poison the body politic.

Whipping up anti-French sentiments, chauvinists started using the term 'Jacobin' to denigrate anybody with radical leanings. British political affairs became polarized, with savage satire an important propaganda weapon on both sides. Caricaturists contrasted stalwart, home-bred John Bulls with under-nourished *sans culottes* (Jacobin workers wore trousers rather than aristocratic knee-length breeches), but they also savaged British politicians, cruelly exaggerating both the short plump figure of Charles James Fox, leader of the Whig opposition, and the pale, ascetic features of the Tory prime minister, William Pitt.

After a couple of false starts, I realized that my poem (*my poem!*—I had already become proprietorial) was not to be found in the *Anti-Jacobin Review*, which was set up in July 1798 with a secret government subsidy, but in its short-lived predecessor, the *Anti-Jacobin; or, Weekly Examiner*, of 1797–8.[25] The original suggestion for this weekly paper came from George Canning, at that stage an ambitious young Tory in the Foreign Office eager to make his mark by defending Pitt, who was currently under fire as a tyrant imposing extortionate taxes to fund the war against France. The journal ran for eight months, right through the parliamentary session for 1797–8, mixing satire, outright ridicule, and straight reporting in its aim to protect the British establishment from subversion. Thirty years later, Canning had himself become prime minister, when he anxiously scrutinized the latest caricatures on display in the windows of London's printshops to make sure that he was still important enough to be mocked.

Recruiting a few friends as editor and contributors, Canning embarked on his project with the zeal of a conspirator. Meeting in secret, the journalists climbed up through a printer's shop in Piccadilly to a secluded first-floor eyrie, even employing an assistant who guarded against their handwriting revealing their identity by copying out all the articles himself before sending them to the press. Crammed together in their small chaotic office, these intelligent, well-educated young men worked through weekends to meet their Sunday deadline, laughing and drinking as they strove to drive home their political points. Like the other *Anti-Jacobin* poems, 'The Loves of the Triangles' was written collectively, drafts being left open on a communal table for anyone to add a few lines whenever they felt inspired. Reading this collaborative verse over two centuries later, I realized how much fun they had, feeding off one another to produce witticisms of escalating hilarity like the running gags on a radio show.

The eight-page paper appeared every Monday, about the size of a modern colour supplement, but printed in black-and-white double columns on flimsy paper. It settled into a fairly regular format. First came the 'Lies, Misrepresentations and Mistakes' perpetuated by supposedly less reliable publications, their falsehoods exposed and exaggerated for *Anti-Jacobin* readers. Next was the week's satirical poem, generally a page or so in length, followed by 'Foreign Intelligence', and sometimes readers' letters (perhaps contributed by the satirists themselves?). Each issue cost sixpence (unhelpfully, the modern equivalent is either £2 or £20 depending on which conversion ratio you use). Like most journals of the time, the *Anti-Jacobin* would mostly have been read by gentlemen in coffee houses and gentlemen's clubs, but its shelf life lasted long after the venture folded in 1798, because the entire run was republished in a smaller format, and the poems were reproduced separately in various editions.

Browsing through my own little book of *Anti-Jacobin* poems (including one in Latin), I realized that 'The Loves of the Triangles' was more wide-ranging than its title implies. It refers explicitly to three of Darwin's works—not only *The Loves of the Plants*, but also *The Economy of Vegetation* and *Zoonomia*, his medical text containing controversial ideas about evolution. I also discovered that this particular parody was unusual in being presented as three separate instalments totalling almost three hundred lines, with substantial footnotes on top of that. This was an encouraging find: political journalists only bother to keep producing such sophisticated jibes when they know they are finding their mark. Darwin was clearly worth targeting—but why?

Poetry and Parody

Wherever you start in the eighteenth century, before long you encounter Alexander Pope. Whether your interests lie in poetry or prose, Shakespeare or the classics, satire or philosophy, Pope inevitably turns up because he was regarded as the Enlightenment expert in all of them. So I was not surprised to find him only a third of the way through the introduction to 'The Loves of the Triangles' (reproduced in the Appendix). The following sentence is supposedly written by a Jacobin called Mr Higgins (patronizingly referred to as Mr H.): 'Our first principle is, then—the reverse of the trite and dull maxim of Pope—"*Whatever is, is right.*"' I recognized the quotation immediately, because so many eighteenth-century authors had chosen it as the epigraph for their books. By far and away their top favourite was the next couplet in the same poem, *An Essay on Man* (1733–4):

> KNOW then thyself, presume not God to scan;
> The proper study of Mankind is Man.[26]

For anyone who relished the security of England's class-structured society, Pope's *Essay on Man* was enormously comforting, and became one of the most influential poems ever written. Although it welcomes the progress achieved through scientific research, it also warns human beings that their knowledge must always be limited, that they should never presume to discern God's intentions. Pope's view of creation is a sophisticated rendition of that British slogan, 'Mustn't grumble'. God, he argues, has excellent reasons for ordering the world as it is and allocating its inhabitants to different roles.

Pope's four-part poem summed up the optimistic philosophy that prevailed throughout Europe in the first half of the eighteenth century. Perhaps inevitably, reaction set in amongst a younger generation, who objected that relying on the wisdom of a Divine Organizer makes it easy to justify oppression and quash improvement. As the *Anti-Jacobin* writers knew, opposition to this self-satisfied conservatism originated in France, where Voltaire's satire *Candide: or, All for the Best* (1759) savaged religious complacency in the face of hardship and disaster. When their fictitious Mr H. retorted that '*Whatever is, is wrong*', he effectively declared himself to be a French-inspired anarchist committed to overthrowing British stability. For those in the know, 'the Author, Mr HIGGINS, of *St. Mary Axe*' was the political reformer William Godwin; although St Mary Axe did exist, very near where the Gherkin stands today, its use as an address conveniently implied that he would import the guillotine and other terrors from revolutionary Paris.

Reluctantly, I realized that to write about Darwin, I had to acquire some technical literary vocabulary. Turning to my poet for advice, I learnt that Pope influenced not only the content but also the style of eighteenth-century poetry. In particular, he made heroic couplets the most fashionable form. The century's top quotation, 'KNOW then thyself, presume not God to scan;/The proper study of Mankind is Man' is a heroic couplet: the final words rhyme, and both lines are written in a metre called iambic pentameter. Derived from Greek, this means that a line has five feet, each one an unstressed syllable followed by a stressed one. Just read the verse aloud, but exaggerate the natural emphasis, and the metre will emerge of its own accord.

Poets think about laws differently from scientists. The physical world never deviates from the mathematical formulae describing it: when apples fall to the ground, they always follow Newton's law. In contrast, poetic metres act as guides reporting how people naturally speak English, which is not with metronomic regularity. This means that not all lines of iambic pentameter have exactly ten syllables: you can have a feminine ending of an extra unstressed syllable—such as 'faster' instead of 'fast'. But by the eighteenth century, the rules had become rigid. Rather like trying to build a curved house out of Lego blocks, adopting formal symmetry makes it hard to express emotion without sounding artificial. With his fine ear and technical skill, Pope excelled at pithy self-contained two-liners. Less gifted poets such as Darwin lacked his ability to prevent that strict structure from being repetitive and boring.

Darwin's style may seem painful, but in his defence, he had little choice about writing in couplets, because that was what most eighteenth-century poets did. By emulating the conventions of the great classical authors—Virgil, Homer, Ovid—they hoped but failed to restore past values and create Britain's own great literary heroes. Instead of producing magnificent epics about nature, they proved better at using satire to describe society.

Pope himself often adopted a mock-epic tone, using a grandiose style to write about trivial topics.

Unlike Pope and other more talented poets, Darwin relied heavily on emotive adjectives and contrived rhymes. Introducing personified flowers in capital letters, he tried to make his poetry playfully erotic, although often it just sounds unintentionally funny. These are some typical lines:

> *Two* gentle shepherds and their sister-wives
> With thee, ANTHOXA! lead ambrosial lives;
> ... With charms despotic fair CHONDRILLA reigns
> O'er the soft hearts of *five* fraternal swains;[27]

When Canning and his friends replaced plants with geometrical shapes, they produced some brilliantly ludicrous scenarios. To heighten their mockery, they exaggerated Darwin's clunky style, and their burlesque highlighted his deficiencies more effectively than any lengthy literary review:

> For *me*, ye CISSOIDS, round my temples bend
> Your wandering Curves; ye CONCHOIDS extend;
> Let playful PENDULES quick vibration feel,
> While silent CYCLOIS rests upon her wheel (14)

Pope tended to convert rivals into enemies by viciously imitating not only their rhyming skills but also their predilection for pedantic footnotes.[28] Similarly, as you can see by looking at the Appendix, the *Anti-Jacobin* writers revelled in parodying Darwin's ponderous annotations. Some of the shorter notes are relatively self-explanatory, although others are delightfully (if frustratingly) convoluted. For example, it took me a long while to solve the mystery of the long pseudo-scholarly monologue explaining why Cinderella was whisked off to the ball by 'six cock-tail'd mice' (line 73). The joke hinges on a complicated pun based on the similarity of the Latin words for 'wall' and 'mouse' in a phrase from Ovid—presumably the subject of many schoolboy howlers in translation lessons.[29]

Although I never did manage to decipher many of the footnotes, chance intervened one morning when an unsolicited catalogue thumped through my letterbox. A spectacular picture caught my attention (Fig. 1). Designed

Figure 1 James Gillray, *NEW MORALITY; – or – The promis'd Installment of the High-Priest of the Theophilanthropes, with the Homage of Leviathan and his Suite* (1798), published by J. Wright for the *Anti-Jacobin Magazine and Review*. © Trustees of the British Museum.

by James Gillray, the caricaturist so vitriolic that victims paid to have plates destroyed, it was included in the farewell issue of the *Anti-Jacobin*, just three months after the third and final installment of the 'Loves of the Triangles'. I had failed to spot it before because I was relying on my reprinted edition of the poems: like working online, convenient but no substitute for looking at originals in a library.

Once installed in the Rare Books Room, I searched out Gillray's large engraving. Bound in with the journal, carefully folded in a concertina to make it fit within the pages, *The New Morality* is produced on such high-quality paper that the sepia lines look like original ink. On the other hand, the letters are wobbly and there are some spelling mistakes, because Gillray had to work in mirror-writing so that his original drawing could be reversed during printing.

The first thing I noticed was the scaly monster to the left with a forked tail and giant paunch, carrying on his back three stout citizens waving *bonnets rouges*, the soft red Phrygian caps worn first by American and then by French Revolutionaries. Next I spotted a giant cap of liberty converted into a 'Cornucopia of Ignorance' disgorging books and newspapers written by so-called Jacobins—William Godwin's *Political Justice*, Mary Wollstonecraft's *Wrongs of Women*, Coleridge's poetry. At the front, Godwin is the pig holding his own book, *Political Justice*, while the wide-jawed reptile weeping crocodile tears and wearing stays represents Tom Paine, the Norfolk corset-maker who became one of the Revolution's staunchest English supporters.[30]

I shifted my gaze to the bouquet behind the horn of plenty, and noticed that the wicker basket was labelled 'Zoonomia, or Jacobin Plants'. No ordinary flowers, they had blossomed into the *bonnets rouges* and tricolour cockades of the French Revolutionaries—and *Zoonomia* was the title of the medical textbook by Darwin. Here was further proof that Darwin was of enduring and central political significance for those who, in this hyper-emotional atmosphere, feared all things French.

Scientific Politics

Flicking through my *Anti-Jacobin* poetry collection, I found some self-explanatory titles—'Sonnet to Liberty', 'Ode to Jacobinism', 'La Sainte

Guillotine'. But 'The Loves of the Triangles. A Mathematical and Philo-sophical Poem'? What did that have to do with Anglo-French affairs? My first thought was of a three-way personal relationship, and Google duly steered me towards some untempting websites. Avoiding those, I discov-ered in the *OED* that triangularity acquired its current amorous spin only towards the end of the nineteenth century. By triangles, the *Anti-Jacobin* meant triangles—geometrical diagrams. But why should they be politi-cally significant? Pasteur's maxim came into play. Chance favours the pre-pared mind—and the unusual name of John Robison nudged me into noticing that this eminent professor of physics at Edinburgh University had published an extraordinary polemical tract in the previous year.

Robison's *Proofs of a Conspiracy* blamed the revolutionary disorder in France on a secret mystical network that had (supposedly) permeated and corrupted Masonic lodges on the continent (British sects were presumed to be untainted). The aim was, he warned, to extinguish Christianity and destroy civil society. He also contributed to a special supplement of the *Encyclopaedia Britannica* designed to counteract 'the seeds of Anarchy and Atheism' spread by 'that pestiferous work', the French *Encyclopédie*. In his chauvinistic articles, Robison presented a specifically British style of carry-ing out research.[31] Although Robison was an extreme example, many of his colleagues automatically distrusted science from across the Channel: for them, how people investigated the world was inseparable from their religious and political beliefs. According to the traditional Christian view, since God had planned the universe in every detail, scientific study was one way of approaching nearer towards Him. It followed that because French radicals denied the existence of God, they were incapable of under-standing His creation.

Ideologically, science was an international subject that knew no bound-aries. In reality, the traditional animosity between France and Britain often spilled over into intellectual territory. On one side of the Channel, Newton reigned as the national hero who championed experiment and observation; on the other were marshalled the followers of his rival, René Descartes, who had favoured rational, deductive logic. London and Paris each had its own Royal Society promoting science, but the two organizations operated very differently. British monarchs lent little more than their official approval, whereas the French state poured money into research. As a consequence, science was conducted very differently in the two countries.

Chemistry in particular came under suspicion in Britain. Many caricatures from this period feature an obsession with gases, especially any exploding from a human orifice. Although scatological jokes always prompt an appreciative snigger, at the end of the eighteenth century almost anything to do with fumes or clouds, balloons or bangs, referred both to chemical experiments and to political insurrection. Paris's most eminent chemist, Antoine Lavoisier, made the links explicit by claiming to have revolutionized chemistry. Insisting on precise measurement and quantitative analysis, Lavoisier invented new Latin labels for familiar substances such as Epsom salts or baking soda. He was also an expert on gunpowder, which made gases an even more appropriate symbol for dangerous French radicalism. And national pride was at stake: Lavoisier gave the name 'oxygen' to a gas that had recently been isolated by one of Darwin's close friends at the Lunar Society, the Birmingham chemist Joseph Priestley.

The most charismatic and stalwart defender of British tradition was Edmund Burke, distinguished Member of Parliament, exceptionally gifted orator, and very effective propagandist for the pro-establishment views of the *Anti-Jacobin*. Over the centuries, argued Burke, countries have developed the systems of law and government that work best for them and should be preserved. He excelled at stimulating reactionary views through exaggerating the risks involved in change. William Stevens, Darwin's nervous neighbour, confided to his diary that 'I for one think the Minister will complete the downfall of the present system and the establishment of a Republic in this Country.... If you believe Burk's [*sic*] statement one fifth of the Community are already incorrigible Jacobins.'[32]

Burke's *Reflections on the French Revolution* (1790) sold thousands of copies. He not only hurled invective at Parisians, but also lambasted opponents at home who disagreed about how the country should be run. One of his main targets was Priestley, who posed a multiple threat: he supported the French Revolution (as did his powerful patron, Lord Shelburne), his followers came from the wrong social class, he was a Unitarian minister (and so rejected conventional belief in the divinity of Christ), and he practised chemistry. In other words, he represented all that was abhorrent to Burke, who—in a fine rhetorical flourish—likened the French spirit of liberty to a wild gas that had broken free from a violent process of effervescence.

For Francophobes, the *Anti-Jacobin*'s triangles were not merely lines drawn on paper: to use mathematical imagery was in itself to make a political statement. Modern science relies on abstract mathematical manipulation, but Robison and his contemporaries were trained in the traditional common-sense approach of analysing only those things that you can see, touch, and measure. They denounced Parisian algebra and calculus as being flowery, superficial, and ostentatious—just like the French themselves. When Cambridge University tried to alter its syllabus by introducing more mathematics, lecturers and students alike insisted that would merely take time away from studying the Bible and the classics. After Napoleon seized power in 1799, the top French students were cherry-picked to design the bridges and roads necessary to maintain military rule. In contrast, British educators regarded engineering as an inferior subject: real gentlemen should work with their brains rather than their hands, and not stoop so low as to expect payment for their work.

British aversion to mathematics was reinforced by associating measurement with officialdom. Uniformed excise men were grading British beer with scientific instruments rather than by its taste, and Burke wrote despairingly that quantifiers had destroyed the age of chivalry. For him, calling someone a calculator was an insult. Measurement was, he pronounced, 'a ridiculous standard of power...and equality in geometry the most unequal of all measures in the distribution of men'. He was referring to the French plan of allocating votes by replacing traditional provinces with a geometrical grid. In a bid for centralized power, he wrote, Paris was setting up a geometric constitution requiring 'Nothing more than an accurate land surveyor, with his chain, sight, and theodolite'.[33]

Significantly, the major land-surveying technique was called triangulation. Although physically demanding, the basic principle is straightforward. A surveyor stands on a hill and looks down at an imaginary triangle spread out beneath him (and they were all men). By measuring the length of its base and two of its angles, he works out the lengths of the other two sides; he then uses one of these to form the base of an adjacent triangle, and repeats the process so that an invisible geometric chain gradually snakes across the countryside.

When 'The Loves of the Triangles' appeared, two French astronomers were just completing a large triangulation project designed to determine the length of the metre by measuring a line of longitude. From a British perspective, this was a deeply misguided project that aroused great antagonism. For one thing, adopting the metre meant rejecting traditional units of measurement such as inches and feet, which are based on human dimensions. Decimalization raised further problems. As well as rationalizing distance, the French had adopted a Republican calendar, in which the seven-day weeks that conform with the Bible had been replaced by ten-day blocks. Scientific debates about the metric system were permeated with national hostilities. Basing the metre on the size of the Earth makes the system sound neutral and international, but French surveyors insisted that one particular line of longitude would work best—one that ran through French territory—and that French instruments were more suitable than British ones. (One of the astronomers was too embarrassed to confess that he had made a mistake, so the standard metre bar carefully preserved in Paris was in fact the wrong length.)

About halfway through 'The Loves of the Triangles', its authors started playing with the number three—perhaps no coincidence as there were three of them, George Canning, John Hookham Frere, and George Ellis. The number's first appearance is in a direct parody of a couplet by Darwin about a flower with three male stamens and one female pistil:

> The freckled IRIS owns a fiercer flame,
> And *three* unjealous husbands wed the dame.[34]

This is the *Anti-Jacobin* version:

> *Three* gentle swains evolve their longing arms,
> And woo the young REPUBLIC's virgin charms (139)

Liberty, equality, fraternity: the revolutionary slogan lent itself to threesomes. One popular motif in France was the Masonic triangle with its all-seeing internal eye, which still appears on American banknotes. The poetic trio continued this theme for over eighty lines—the three witches in *Macbeth*, the three fates, Lear's three daughters.

The authors ended by describing a hectic downhill carriage ride with three passengers squashed together in the Derby dilly at Ashbourn (see lines 178–83 in the Appendix). I'm still waiting for the serendipitous opportunity to find out who they were.

CHAPTER THREE

A Triangle of Poets

I, too, dislike it: there are things that are important beyond all this fiddle.
Reading it, however, with a perfect contempt for it, one discovers in
it after all, a place for the genuine.

Marianne Moore, *Poetry* (1921)

E rasmus Darwin was not the sort of man to worry overmuch about
literary fashions. He knew what he liked—'that Sort of Poetry
which outflamed the boundary of Nature'. Of his own poems,
two lines were said to be 'his peculiar Favorites.—"To Etna Hecla calls/
And Andes answers from his beaconed Walls." "This might" he said "be
Bombast but it pleased him." '[35]

Darwin's impressive sales figures suggest that many people shared his
pleasure in 'Bombast'. I am not one of them. However intellectually
fascinating his theories may be, I find it hard to avoid wincing at lines
like these, which are taken from his first long poem, *The Loves of the
Plants*:

> AGAIN the Goddess strikes the golden lyre,
> And tunes to wilder notes the warbling wire...
> The Nymph, GOSSYPIA, treads the velvet sod,
> And warms with rosy smiles the watery God;[36]

Putting my personal aversion on one side, I asked myself *why* so many
people bought and admired his books. To start with, the cultural climate
was very different then from now. Poetry was popular, and unlike the vast
majority of modern poets, Darwin sold so many books that he made a
substantial profit. Reviewers enthused about 'the splendour of his poem,

which could not have been more highly finished, sweeter in the flow of its numbers, more exquisite in the expression, more diversified in the matter, or richer in every species of embellishment'.[37] Even the Romantic poets who later came to revile Darwin did at least acknowledge his significance. Looking back at his own youth, Wordsworth remarked that 'my taste and natural tendencies were under an injurious influence from the dazzling manner of Darwin'.[38]

Fashion changes in literature as in everything else. Darwin's lines grate on modern ears, but his style was far from unique in the later eighteenth century. However florid it might now sound, his poetry was not singularly over-the-top. When the political radical William Godwin urged the twenty-year-old Percy Shelley to read the literary greats of the past— Shakespeare, Milton, Bacon—he bracketed Darwin with several other unsuitable influences of the time: '*You* have what appears to me a false taste in poetry,' he admonished his future son-in-law; 'You love a perpetual sparkle and glittering, such as are to be found in Darwin, and Southey, and Scott, and Campbell.'[39]

Because of his success, Darwin constituted an ideal target for aspiring authors trying to promote their own literary ambitions. By criticizing Darwin, the next generation of poets aimed to forge their own identity by reacting against the floridity of their elders. They regarded him not so much as an individually poor writer, but more as their chosen representative of an outdated poetic movement—those weak imitators of Pope who could contrive only ornate and overblown expressions. Coleridge joined in the attack, stating bluntly: 'I absolutely nauseate Darwin's poems.'[40] Similarly, Lord Byron parodied Darwin's leaden scansion and contrived rhymes, deliberately writing like an old-fashioned poet:

> But not in flimsy Darwin's pompous chime,
> That mighty master of unmeaning rhyme,
> Whose gilded cymbals, more adorn'd than clear,
> The eye delighted, but fatigued the ear;[41]

Darwin insisted that poets should versify only what they can see in front of them, visualizing it vividly so that readers are transported to the scene in their mind's eye. One fan enthused about the vicarious joy of shivering with his snowdrops, hearing the sighs of his lovesick violets, and watching his jealous cowslips hang their heads.[42] In contrast, the Romantics deliberately turned their gaze inwards, emphasizing the importance of imagination and self-reflection. In the preface to *Lyrical Ballads* (1798), Wordsworth insisted that poets should give expression to their inner feelings. Railing against formalized, formulaic poetry, he recommended using ordinary words that have a function, rather than being decorative.[43] One reviewer described this poetic antagonism as a war being waged between 'the Lake and Darwinian schools of poetry', explaining that they 'are the very antipodes of each other—hostile in all their doctrines, and opposite in every characteristic'.[44]

I turned to thinking about Darwin's voluminous notes. The playwright Noel Coward once remarked that having to absorb a footnote at the bottom of a page is like going downstairs to answer the doorbell when in the middle of making love.[45] Apparently eighteenth-century readers liked being interrupted—once I started looking, I soon found plenty of other poems that were also heavily annotated. This explicit didacticism was not an Enlightenment innovation, but a tradition inherited from classical ancestors, such as Virgil and Lucretius. Because of its memorable rhymes and rhythms, poetry is relatively easy to learn by heart, which makes it a valuable oral medium for teaching. One of Darwin's readers explained the superior effectiveness of Linnaean poetry:

> But then in prose Linnaeus tattles
> And soon forgot is all he tattles.
> While memory better pleas'd retains,
> The frolicks of poetic brains.[46]

In several ways, Darwin's poems resemble William Falconer's *The Shipwreck* (1762), also an extremely popular narrative poem carrying lengthy notes. Myself, I find this heart-lurching account of a sailor surrounded by dying comrades even less appealing than *The Botanic Garden*. But then I'm not living in a period when the Royal Navy was the world's largest management organization, when ships were vital for trade and defence, and when—as Samuel Johnson put it—travelling by sea was so dangerous that it was like being in prison but with the added probability of being drowned.

Recommended as a valuable teaching aid for naval neophytes, Falconer's poem was repeatedly republished into the nineteenth century, often with lavish engravings. Like Darwin, Falconer wrote in heroic couplets, included notes explaining technical terms (such as mizzen, clue-garnetts, and thrott), and sprinkled his verse with mythological characters (always in capital letters). Although he leaves out the sex, Falconer writes in the same didactic vein as Darwin; in this extract, he is confident that his readers will immediately recognize Phoebus as the Sun:

> The Pilots now an Azimuth attend,
> On which all courses, duly shap'd, depend:
> The compass and the octant ready lay,
> Reflecting planes and incidents survey;
> Along the arch the gradual index slides,
> While PHOEBUS down the vertic' circle glides...[47]

Reading poems like Falconer's makes Darwin's own seem less extraordinary. Viewed in retrospect, such authors exhibit Janus-like features. By celebrating the latest inventions, they seem to be looking towards the future; yet poetically, they obstinately cling to the past, retaining an old-fashioned style that antagonized younger writers. This division reflects how Darwin is now perceived by academics. Historians of science are impressed by his scientific expertise and his apparent prescience of modern technology, but read his poems only to laugh or to scour them for evidence. Reciprocally, those experts in English literature who have examined Darwin's impact on the Romantic poets mostly care little for his ideas about evolution, canals or steam engines.

Because of this disciplinary ring-fencing, many months went by before I came across *Romantic Atheism* by Martin Priestman, a professor of English and one of the few modern scholars to take all aspects of Darwin's work seriously. I had long been aware of the bonds linking Darwin to contemporary inventors, doctors, and scientific industrialists. Later, after I began this project, I discovered to my surprise—as I have just described—that his type of poetry was not unique or even exceptional in the eighteenth century. After encountering Priestman's book, I headed off in yet another direction, exploring at a more abstract level how Darwin's ideas corresponded to those of other Enlightenment poets.

In *Romantic Atheism*, Priestman brackets together Darwin and two other eighteenth-century poets: William Jones and Richard Payne Knight.[48] At first, I was perplexed by Priestman's identification of this poetic trio, because they are remembered—if at all—in very different ways. As thumbnail sketches, Darwin is characterized as a provincial physician and botanist embedded in an industrial network, Knight as a wealthy antiquarian and landscape designer, and Jones as a judge in India who stimulated British orientalism. What, I wondered, could these three men have to do with one another?

Some relationships were easy to spot. For one thing, they directly affected each other: Jones's early poems influenced Darwin's *Botanic Garden*, which in turn influenced Knight's *The Landscape; A Didactic Poem* (1794). Contemporaries also grouped them together, although generally in pairs rather than as a threesome—and not always in flattering terms. Politically, the three men were loosely aligned in the opposite camp from Canning's satirists. As a Member of Parliament, Knight opposed Pitt's regime; Jones and Darwin supported the American and the French Revolutions, and both were friendly with Benjamin Franklin. Most relevantly for me, the *Anti-Jacobin* aimed almost simultaneously at Knight and Darwin.

But were there deeper resemblances between these three authors? I realized that because I knew what had happened over the intervening two centuries, my preconceptions were clouding my perceptions of these writers. Like organizing a pack of cards, there are various criteria for classifying people and sorting them into groups. The modern labels for their activities have slotted these three men into three mutually exclusive categories: science, aesthetics, and linguistics. In contrast, their colleagues assessed them not by their lasting achievements, but by their daily activities and behaviour. Gradually, I came to perceive the significance of their shared progressive approach towards nature and towards society. The *Anti-Jacobin* catered for reactionary readers who feared the unknown, who preached that any change would inevitably make things worse. In contrast, these three idealists envisaged improvement towards a natural state of perfection.

But before reaching that conclusion, I learnt more about Jones and Knight. So for readers as bewildered as I was initially—I hope these summaries will help.

William Jones (1746–1794)

Once I had sorted out which of the many eighteenth-century William Joneses I was pursuing (one namesake was his own father, an eminent Welsh mathematician and friend of Isaac Newton), I realized that I had encountered him before. Or at least, I knew about his famous follower, the bestselling Victorian poet Edward Fitzgerald, whose translation of *The Rubáiyât of Omar Khayyâm* (1859) is still many British people's first encounter with Eastern philosophy. Its most evocative lines—'The moving finger writes; and, having writ,/Moves on' or "Tis all a chequerboard of nights and days'—probably owe more to Fitzgerald than to the Persian original he was working from, but his main scholarly guide was Jones, who wrote grammatical texts as well as translating Middle Eastern literature.

As a teenager, Jones had plenty of influential connections, but no father and no money.[49] He did, however, have an enlightened mother. Even though she was banned from going to university herself, she insisted that the only way to succeed in life is by reading. By the time he was twenty, Jones had read so much and knew so many languages that his Oxford college made him a Fellow before he had graduated. Within another six years, he was a Fellow of the Royal Society, belonged to Samuel Johnson's circle of close friends, and had published several books.

Jones's career was marked by contradictions. A keen political radical who deplored aristocratic privilege, his first published translation was the sympathetic biography of a Persian despot. After switching to law so that he could earn his own living, Jones ended up as a colonial administrator in India, relying on influential backers for recommendations and helping his own protégés to gain promotion. A devout Christian, Jones embarked on a mission to promote Asian culture that involved him in sacrilege twice over—Hindu poetry celebrates multiple gods and was, by European standards of the time, immoral verging on pornographic.

Ironically, although he accused his colleagues of denigrating any mode of life except their own, Jones's monument at Oxford shows three Asian scholars sitting at his feet as if absorbing his words of European wisdom. This imperialistic configuration ignores his lifelong project to reverse traditional roles by learning about other civilizations instead of unthinkingly exporting Western ideas. Jones was atypically sensitive to the customs of others. In a letter to Burke, at that stage a close friend, Jones emphasized how perversely tyrannical it would be to impose their home system of justice on a country that had already developed its own governmental regime. He lived trapped in a dilemma that still troubles liberals: does giving priority to freedom of choice entail tolerating oppressive practices elsewhere?

Although Jones could have practised law in London or Oxford, he decided to work in Wales, where his grandfather had been a sheep-farmer. Trying to live up to his high ideals, Jones acted for his poorer clients free of charge, defending local Welsh workers against their exploitative English landlords and administrators. Paralleling his later activities in India, Jones also resurrected Celtic customs, initiating druidical meetings and writing patriotic poetry. He later moved back to London and campaigned for political reform, making himself unpopular by opposing the government's policy in India, attacking the slave trade, and supporting universal suffrage (by which, like everybody else then, he meant votes for men).

Eventually, in his late thirties, Jones landed the well-paid job he had coveted for years—High Court judge in Calcutta. Persevering in India's unforgiving climate until he died eleven years later, Jones ruined his health through overwork as he struggled to reconcile British rule with local practices. When tensions escalated between Burke and Warren Hastings, governor-general of Bengal, about the form that British intervention should take, Jones was caught in the middle and handled the situation badly. Burke and Jones were never friendly again.

When he arrived in India, Jones's first step had been to establish the Asiatick Society of Bengal. This unique institution came to resemble a remote outpost of the Royal Society, its members mainly upwardly mobile British employees of the East India Company. Jones's initiative encouraged them not only to learn more about their new home, but also to make people back in Britain aware of Asian culture by producing a scholarly journal, *Asiatick Researches*. Jones hoped that if British administrators understood more about India, they would govern the country more fairly and more effectively.

As the Society's president, Jones contributed over a third of the articles in *Asiatick Researches*, which was distributed (and plagiarized) all over Europe. He covered an extraordinary range of topics, and his enthusiasm for all things Asian shines out from its pages. Jones's energy was stunning. On top of heavy sessions in court and long hours devoted to codifying Hindu and Muslim law, he wrote with great erudition about archaeology, botany, music, physiology, religion, and more. His publishing venture was well received—'How grand and stupendous,' one reviewer enthused; '....We may reasonably expect to enlarge our stock of poetical imagery, as well as of history, from the labours of the Asiatick Society...to combine the useful and the pleasing'.[50]

'Oriental' Jones worked hard to earn his nickname, praising the sophistication of Arabic, Persian, and Turkish cultures in book after book. He remembered his initial encounter with Sanskrit literature as a revelation, an eye-opening experience even more moving than reading the *Iliad* for the first time. To help other Europeans share the delights he had discovered, in the very first volume of *Asiatick Researches* Jones proposed a technique for transcribing Eastern languages that remains influential even today. Through his numerous translations, essays, and original poems, Jones convinced Europeans that Eastern literature should rank as high as the Greek and Roman classics. Among his great international successes was *Sacontalá* (1789), his version of a play about a semi-divine courtesan who captivates a king with her beauty. With its unfamiliar yet alluring blend of the sacred and the profane, this erotic drama became a European emblem of Indian civilization. Jones influenced the Romantics—especially Coleridge and Byron—as well as poets of the future, including not only Fitzgerald, but also Alfred Tennyson and T. S. Eliot.

Jones's investigations into botany and medicine are less well-known than his work on law and languages, but they are pervaded by the same ethos. They reveal a man who loved to organize and to classify, but who was also determined not to impose Western systems and values on to the

Asian cultures that he admired so much. He and his wife both loved collecting rare plants, and his report in a letter that 'Anna has brought me a
new flower, and we are going to hunt for it in Linnaeus' confirms his allegiance to the latest European taxonomy.[51] Nevertheless, Jones also sought
out local knowledge from indigenous informants, and his learned essay on
Indian botany is packed with references to the Sanskrit names of plants
and their significance for Hindus.

Amongst modern academics, Jones is now most celebrated for founding
the new discipline of historical linguistics. His work was revolutionary both
because of the techniques he developed and because of the conclusions he
drew. As a doctor, Darwin studied evolution in the biological world, making suggestions that would still seem shocking more than half a century
later. As a pioneer in linguistics, Jones brought the same approach to human
civilization, amassing evidence to show that different modern languages
have descended from a single ancestor. Or in the words of a British doctor
and botanical enthusiast writing from St Petersburg, he 'demonstrates the
common Origin of the European Nations and Languages, and that the
Grecian Mythology and ingenious fables, are Still found in the Sanscrit
Books of india, and the Grecian Dieties in the Pagodas of that Country'
(the erratic spelling and capitalizations are his, not mine).[52]

Like a scientific experimenter, instead of speculating abstractly about
languages, Jones compared words and grammatical structures to work out
how they had originated. In a way, learning how to speak a language
resembles developing a botanical system of classification, because describing the world involves grouping similar things together and using related
words to label them. And like plants and animals, languages change or
evolve over long periods of time. By meticulously analysing Greek, Latin,
and Sanskrit, Jones found that the elements they had in common were far
too numerous to be accounted for by accident—these three languages
must, he insisted, have branched out from a single common source, an
older language that perhaps no longer existed.

Jones died four years before Knight and Darwin were savaged by the
Anti-Jacobin, but he shared many of their political ideals and had been
notorious as an opponent of the Tory government. At the first meeting of
the Asiatick Society, Jones suggested that Indians should be included as

members. Although this revolutionary recommendation was never adopted, *Asiatick Researches* became the first journal produced by Europeans to carry articles by Indian scholars. Canning and his fellow Pittite satirists would not have approved of Jones's anti-establishment, pro-Indian initiatives.

Richard Payne Knight (1751–1824)

While Jones was overworking himself towards an early death in a tropical climate, dreaming in vain of retirement to a comfortable home in Britain, Knight was living on the edge of Wales, lavishly spending the fortune he had inherited from his grandfather's ironworking business.[53] Lavishly but not extravagantly: perhaps because of his industrial origins, Knight made better use of his money than those aristocrats who worked their way in a decade through wealth that had been carefully husbanded over generations.

Like both Darwin and Jones, Knight was an inconsistent radical. He repeatedly antagonized the establishment aristocrats with whom he mingled by contrasting the freedom and rationality of ancient Greece with the oppressive regime of the British monarchy and Church, yet his political protests sat uncomfortably against his elitist aesthetics and his penchant for costly luxuries. As an outspoken supporter of the French Revolution and an MP who opposed Pitt, Knight was an ideal victim for the *Anti-Jacobin* writers.

Instead of going to university, Knight travelled to France and Italy on a Grand Tour, a common way for young gentlemen to complete their education. Unlike most of his companions, he excelled at Greek and was genuinely fascinated by architecture. As soon as he got back, Knight began using what he had learnt to indulge his expensive tastes. His first project was to build a house—Downton Castle—on his Herefordshire estate. Although work was intermittently interrupted by trips abroad and to London, the large house dominated the rest of his life. Eclectic in style (a combination of Welsh parapets and the Roman Pantheon), Knight's neo-medieval construction launched a British fashion for castellated buildings set in a carefully designed landscape.

Knight converted the grounds of Downton Castle into a massive and expensive experiment, testing different ways of remodelling the landscape in the new picturesque style. 'Picturesque' originally meant 'fit for a picture', and Knight taught that garden scenes should be modelled on pictures and obey the 'three distances' rule of landscape painting. He revamped his own garden to make it resemble a painting by the famous seventeenth-century French artist, Claude Lorrain. This was cutting-edge design. Knight and other pioneers of the picturesque movement—notably William Gilpin, a schoolteacher and clergyman—were out to replace the sense of rolling tranquillity induced by the previous generation of gardeners. Whereas Capability Brown and his contemporaries had aimed to make artificial views look more natural than nature herself, the latest vogue was for rugged inhospitable landscapes, their desolation enhanced by prefabricated Gothic ruins carefully placed at apparently random spots.

Tourists were abandoning their traditional pilgrimages to the great classical buildings of Italy, and instead were venturing into the Alps and remote parts of the British countryside, where they sought out the sense of exhilarating fear gained by peering over the edge of a precipitous gorge or being deafened by tumultuous waterfalls. When the poet Thomas Gray visited the Scottish Highlands, he exclaimed: 'The mountains are ecstatic....A fig for your poets, painters, gardners [sic] and clergymen that have not been among them; their imaginations can be made up of nothing but bowling-greens, flowering shrubs, horse-ponds, Fleet-ditches, shell grottoes and Chinese rails.'[54]

The guidelines for these artistically minded travellers had been laid down by Edmund Burke. Long before he became a prominent politician, Burke published the century's most influential book on aesthetics: *A Philosophical Inquiry into the Origin of our Ideas on the Sublime and Beautiful* (1757). He believed that contemplating a landscape was not an intellectual activity, but instead entailed emotional immersion. An engaged spectator would be overwhelmed by a gut reaction to either its pleasing harmony or its awesome grandeur—or in Burke's terminology, to its beauty or its sublimity.

Burke's binary opposition affected how travellers regarded the scenery around them. When the Enlightenment reformer Arthur Young visited the new Midlands factories, he was struck not by their ugliness or squalor, but by their sublime contrast with the area's natural beauty. 'Coalbrook-dale itself is a very romantic spot,' he reported, 'too beautiful to be much in unison with…the noise of the forges, mills, &c. with all their vast machinery, the flames bursting from the furnaces with the burning of the coal and the smoak of the lime kilns, are altogether sublime, and would unite well with craggy and bare rocks….'[55]

The picturesque was introduced as a third and intermediate aesthetic category by Gilpin and other landscape specialists. In their instruction manuals, they taught gardeners to create appealing scenes with winding paths and asymmetrical buildings, and it was under their guidance that artists bought special viewing glasses to transform the scene around them into a pleasing composition for a painting. When the first British settlers depicted Australia, they conformed to such expectations by giving sinuous undulations to the ramrod-straight eucalyptus trees.[56]

There are gardeners and then there are garden designers. Knight was less interested in practical details than in reflecting on the philosophical aspects of aesthetic pleasure. One portrait shows him surrounded by valuable objects that he is ignoring in favour of reading about them in his book. Knight borrowed from the Scottish philosopher David Hume to insist that beauty is not an intrinsic quality, but lies in the eye of the beholder. For him, unlike for Burke, admiring a landscape was an intellectual as well as an emotional process, because what you see evokes memories in your mind—and your personal mental associations will not be the same as those of the observer standing next to you. A broken-down hovel is not inherently picturesque, Knight argued, but only carries emotive power because it summons up previous pictures and ideas.

Because artistic appreciation involved the mind as well as the feelings, judgement could be acquired—and if taste could be taught, then it could also be bought. Following Knight's principles, rich and well-educated collectors—in other words, men like him—were able to cultivate the discernment required of a connoisseur. Internationally renowned for his expertise, Knight belonged to the Society of Dilettanti, an exclusive club for wealthy, convivial gentlemen who gratified their pleasures not only with fine food and expensive wines, but also by buying antiques, pictures, and natural curiosities. Although easy to mock (and they often were), the

Dilettanti were also serious scholars. When asked about Egyptian anthropology, the eminent German naturalist Johan Blumenbach recommended looking 'either in Sir William Hamilton's collections in the British Museum, or in those of your great Antiquarians Mr Townley & Mr Knight'. All three were Dilettanti.[57]

Knight, the would-be dictator of polite taste, marred his reputation by committing two great blunders. The very first book he published claimed to be a learned examination of early religions, and it explicitly acknowledged the well-respected research of William Jones. It provoked uproar. And the title suggests why—*An Account of the Remains of the Worship of Priapus* (1786). For his fascinated yet appalled readers, Knight revealed that some Catholic ceremonies featured wax phalluses (Priapos/us was the Greek/Roman god of sex and fertility). Supporting his arguments with (very) explicit illustrations, he suggested that the Christian cross originated as a phallic symbol, and that all religions are fundamentally based on sexual imagery. This presented a perfect media opportunity for satirists, including those of the *Anti-Jacobin*.

Knight's second spectacular error of judgement occurred when what were later known as the Elgin marbles arrived in Britain from the Parthenon. Without inspecting them for himself, Knight pronounced that they were merely inferior Roman copies. One of his major opponents was the painter Benjamin Robert Haydon, who had campaigned to purchase the sculptures. In a fit of rage, Haydon predicted that while the marbles' fame would rise, Knight would sink into oblivion.

Even though Knight was a privileged member of the nation's intellectual and artistic elite, Haydon was proved right. The British Museum inherited the valuable collection that Knight had accumulated, but few of its present visitors have even heard of him.

Triangular Lives

Reading something for the first time is by necessity a one-off event, a psychological experience that—unlike a laboratory experiment—cannot be repeated. Although you never recapture any initial shiver of delight, this uniqueness also brings positive consequences. The very act of reading transforms your understanding, so that whether minutes or years go by before you next encounter the same text, the second time round you will appreciate aspects that you failed to notice earlier.

Historians proceed heuristically, oscillating between reading, writing, and thinking. When you start out, you have little idea which path you will eventually tunnel out through a confusing mass of detailed information. It is only when you try to explain what you have already discovered that you come to realize where the gaps lie and where you need to probe further. By the time you have erratically reached the end—or to be more accurate, one possible end of many—you have become another person.

As I gradually uncovered links between these three men—Darwin, Jones, and Knight—I realized how difficult it would be to impose any narrative order on their intertwined trajectories through life. How could I present my tangled findings in a logical sequence? Every time I started, I stumbled across a fresh sidetrack to explore. But I have to begin somewhere...

My first choice for a starting point is their shared obsession with gardens and flowers. Knight was a leading authority on landscape design, while both Darwin and Jones wrote botanic poetry. In addition, Knight's younger brother, Thomas Andrew Knight, was a keen botanist and Fellow of the Royal Society—as were Darwin and Jones, who both published scholarly works on Linnaean classification. All three belonged to the same school of thought as Horace Walpole, the grand arbiter of eighteenth-century style: 'Poetry, Painting & Gardening, or the Science of Landscape, will forever by men of Taste be deemed Three Sisters, or the *Three New Graces* who dress and adorn Nature.'[58]

For them, reading poetry should be a visual experience. Darwin invited his readers to 'view the wonders of my INCHANTED GARDEN' as if they were projected onto a screen. His verses were, Darwin wrote, to be contemplated 'as diverse little pictures suspended over the chimney of a Lady's dressing-room, *connected only by a slight festoon of ribbons*'.[59] Extolling Jones, the bluestocking Elizabeth Montagu wrote: 'The descriptions are so fine, & all the objects so brilliant, that the sense aches at them!'[60] Very similar praise was heaped on Darwin by the poet William Cowper: 'What we have seen he makes us see more distinctly, and he renders familiar to us what we have never seen.'[61]

The three poets might or might not have met each other, but they definitely all knew a fourth botanical expert and occasional versifier: Joseph Banks, president of the Royal Society. Perhaps this surveying metaphor is over-contrived, but I shall use Banks as an external triangulation point for exploring connections between my three poets.

Jones and Banks had known each other since childhood. Although Banks was three years older, they both went to Harrow School, and as adults, moved in the same London literary circles: their mutual friends included Samuel Johnson, the artist Joshua Reynolds, and the writer Fanny Burney. They also met frequently at the Royal Society, often dining together at the same clubs with the same people.

After Jones went to India, Banks took advantage of having this reliable colleague installed in such a remote place, and they exchanged chatty letters about disseminating scientific information or chivvying mutual acquaintances to return books they had borrowed. The two separated friends negotiated at length about the best ways of stocking botanical gardens, and Jones obligingly sent back plant specimens, reports on local herbal medicines, and seeds for growing crops such as fine Dacca cotton. As his interest in botany grew, Jones embarked on an ambitious project to classify Indian plants by incorporating local expertise into Linnaeus's scheme.

Jones's retirement plans make poignant reading. After asking Banks to look out for a nearby country house with sufficient grounds for his cattle and his garden, Jones wrote: 'I shall be happy in being your neighbor [sic]; and, though I write little now, will then talk as much as you please.'[62] But Jones never saw England again: three years later, he was dead.

In contrast, Banks's relationship with Knight seems to have been less cordial, although they must have met regularly during Knight's visits to London. They both owned homes in Soho Square, they were both Dilettanti, and Knight knew Banks and Jones sufficiently well to publish correspondence between them in *Priapus*. Moreover, Banks was a close friend of Knight's younger brother Thomas, the botanist.[63] Many letters have survived, and they hint at polite iciness between Knight senior and Banks, who reported that they 'talked over their diversities of opinion in as pleasant terms as we could have conversed on matters in which we intirely [sic] agreed.'[64] They may have been continuing an earlier argument about

Knight's contentious evolutionary ideas, which led Banks to expostulate that his 'strong mind has so much exhausted itself on Greek &c as to leave little force left for the explanation of Natural Phenomena'.[65]

Banks's friendship with Darwin was different yet again. They met during Darwin's six-week visit to London with his second bride (not the only scientific wife to find her honeymoon turning into a business trip) and they subsequently exchanged friendly letters, especially after Darwin embarked on his translation of Linnaeus' *A System of Vegetation*. Although Banks was twelve years younger, he came from a higher social class and wielded far more influence. Even so, at this time their positions were curiously parallel. President of London's Royal Society, Banks ruled over an international network of botanists and imperial developers who were changing the entire world. Operating similarly but on a local scale, Darwin dominated the intellectual life of Lichfield, providing a central linchpin for his colleagues at the Lunar Society who were helping to transform Britain into Europe's first industrial nation.

A second human vantage point for inspecting these three men is their major literary hero, Lucretius, to whom they all explicitly paid tribute. A reviewer's comment on Darwin could equally well have been written about Jones or Knight: 'Dr Darwin, like Lucretius, has endeavoured to blend in his poetical works the grave features of philosophy with the mutable graces and smiling charms of imagination.'[66] When I first came across that remark, I had only a flimsy idea of what it meant, but it prompted me to find out more about Lucretius.

Lucretius (*c*.99–*c*.55 BCE) was, I discovered, a Roman poetic philosopher, now little known but enormously important for English science and literature. Lucretius was emulated not only by Knight, Jones, and Darwin, but also by poets as eminent as Edmund Spenser, John Milton, and Alfred Tennyson. A contemporary of Cicero and Julius Caesar, he left only one surviving work: a long Latin poem called *De Rerum Natura* (*On the Nature of Things*). As the title suggests, it ranges over just about everything: how the universe was created, how it works, and how people should live. Revived in the mid-seventeenth century, the poem's first full translation into English (by Thomas Creech) appeared in 1682, and was republished throughout the eighteenth century.

Lucretius' *De Rerum Natura* met conflicting responses in England. Many people admired his attempt to make science palatable by putting it into verse, and his descriptions of swirling atoms corresponded to the latest physical theories. On the other hand, orthodox Christians rejected Lucretius' fundamental philosophy. According to him, the cosmos was created as the result of random events; according to them, to dispense with God's involvement is to undermine morality. For reactionaries, Lucretius symbolized the dissolution of old ways of thought in the face of modernity.

Despite the multiplicity of competing theories, when you get down to it there are fundamentally only two ways of thinking about ordinary matter. Either it is continuous—in which case, the universe is packed full of stuff, with no gaps anywhere; or it is discrete—self-contained particles are separated by empty spaces between them. Before the seventeenth century, most European philosophers belonged to the continuity school of thought, and followed some version of Aristotelianism. Opinion gradually shifted to the discrete camp in which Lucretius had belonged, and he came to be associated with leading atomists such as René Descartes and Isaac Newton.[67]

Lucretius' moral messages had originated a couple of centuries earlier with the Greek philosopher Epicurus (341–270 BCE), who emphasized the importance of searching for personal happiness. This dictum was often misinterpreted. By saying we should maximize our own satisfaction, Epicurus meant not that we should party the whole time, but that we should aim for health, intellectual fulfilment, and emotional equilibrium. Lucretius borrowed this doctrine as the basis for *De Rerum Natura*, which aims to persuade a stressed politician that he should change his hectic lifestyle and move to a quiet country retreat.

What made Epicurus seem sacrilegious to European Christians was his insistence that our present existence is all we have. In his cosmos, there is no life after death, and no God the Great Designer—everything has arisen from the chance encounters of atoms colliding as they move through empty space. Translators defended themselves by insisting that although they admired Lucretius' poetic abilities, they certainly did not endorse his religious views. After publishing only one volume, the diarist John Evelyn became so concerned about the poem's demoralizing effects that he relegated the unfinished manuscript to his study. Creech, a young Oxford don in his early twenties, did at least reach the end, but he vindicated himself by declaring that 'the best method to overthrow the *Epicurean Hypothesis*...is to expose a full System of it to publick view'.[68]

When Newton was dubbed a Lucretian, he resolved the potential conflict between his own Christianity and his allegiance to atomism by asserting that 'the philosophy of Epicurus and Lucretius is true and old, but was wrongly interpreted by the ancients as atheism'.[69] Nevertheless, in the mid-eighteenth century, the Jesuits felt so threatened by the implications of Newtonianism that a French cardinal published *Anti-Lucretius* (1747), a retaliatory poem written in Latin—but soon published in English as a defence against atheism and immorality. And the translator? He was the father of George Canning, co-author of the *Anti-Jacobin* attacks on Knight and Darwin.

By the late eighteenth century, emulating Lucretius had become acceptable. In 1805 another full translation appeared, this one by John Mason Good, a physician who singled out Jones as a modern Lucretian and repeatedly quoted both Jones and Darwin approvingly in the footnotes.[70] Good tried to blunt the poem's atheistic implications by arguing that because the Greeks and Romans lived before Christ, they had not had the good fortune to learn about God through the Resurrection. Whatever one might think about the validity of that argument, its articulation confirms that theology mattered. French prophets of Enlightenment denounced Christianity as superstition, but in Britain, God had not disappeared from scientific explanation.

In terms of their attitude towards Christianity, for once the three poets can be arranged along a straight line: Jones was the most orthodox, Knight lay furthest towards religious scepticism, and Darwin fell somewhere in between.[71] Whereas Newton himself had believed in an interventionist God who is immanent throughout the universe, many of his followers—such as Darwin—adopted various versions of deism. Although deists believed in God, and often regularly attended church, they reduced His role, regarding Him rather like a distant landlord. For them, God is some sort of external entity with no power to interfere in the world, which runs independently and is ruled by the laws of nature. For more conventional Christians, deism was but a polite name for atheism—and from there, it was easy to accuse its adherents of supporting French radicalism. Indeed, this was a step that Knight actually took. In his long poem on landscape design, Knight blamed the Christian Church for the decline of art and knowledge since classical times, and likened the French Revolution to an overflowing stagnant pond that initially causes great damage, but eventually fertilizes the land it has flooded.

Rather than regarding Christianity as a unique civilizing force, all three looked further back in time. Most influentially, Jones focused on Middle Eastern culture, claiming that Greek, Latin, and Sanskrit had all originated from an earlier language. Knight placed the origins of Christian rituals in ancient Egypt and the classical world, while Darwin suggested that Old Testament stories had originated in Egyptian hieroglyphs and were transmitted by Moses.

Because Hinduism lay outside the Judaeo-Christian tradition, it was denounced in Britain as paganism, which was often regarded as close to savagery. Jones cleverly made Hinduism appear civilized by comparing it with the 'Pagan Theology of old *Greece* and *Rome*'—hard to accuse the classical world of being primitive. Giving his arguments a patriotic flavour, Jones made Indians sound hard-working and hence like Europeans: Hinduism is, he pointed out, 'devoutly believed by many millions, whose industry adds to the revenue of *Britain*'.[72]

Not everybody agreed that Jones had successfully reconciled West and East. For example, one of Darwin's Lunar colleagues, the chemist Joseph Priestley, criticized Jones's tolerant attitude towards Indian sexual practices and accused him of deviating from the straight path of Christianity. There were also political implications. Like Jones, French deists emphasized the virtues of paganism, but their aim was very different: they wanted to undermine orthodox Christianity. As a consequence, Hinduism became associated with political radicalism. One of Jones's protégés, Thomas Maurice, found it hard to reconcile his passion for India, his Christian faith, and his support for the British anti-Jacobins. Hinduism, he wrote, was 'the *debateable* [sic] *ground* on which the fury of Jacobin hostility had reared her most triumphant banners'.[73]

In their poetry, all three writers adapted Lucretius' Roman polytheism by deploying gods as allegories for natural forces—classical ones for

Darwin and Knight, Hindu for Jones. Crucially, they envisaged a quasi-sexual energy as both the creator and motivator of nature. Sexual activity was to be enjoyed, to be celebrated for its role in perpetuating life and transmitting improvements to subsequent generations. Darwin, for example, not only anthropomorphized the sex life of plants and animals, but also personified Nature as a generative sustaining goddess:

> IMMORTAL LOVE! who ere the morn of Time,
> On wings outstretch'd, o'er Chaos hung sublime;
> Warm'd into life the bursting egg of Night,
> And gave young Nature to admiring Light!—
> You! whose wide arms, in soft embraces hurl'd
> Round the vast frame, connect the whirling world![74]

In writing these lines, he was closely following Lucretius, who introduced his own poem by addressing Venus as 'the guiding power of the universe', the progenitor of a world teeming with life. '[I]nto the breasts of one and all,' runs a prose translation, 'you instill alluring love, so that with passionate longing they reproduce their several breeds.'[75]

Darwin's adaptation might now be condemned as plagiarism, but to imitate was then seen as paying tribute. In his version of creation, Jones also starts from Lucretius, but renders Venus as the goddess Bhavani, who brings the world to life through the power of sexual love:

> When time was drown'd in sacred sleep,
> And raven darkness brooded o'er the deep...
> The forms of animated nature lay;
> Till o'er the wild abyss, where love
> Sat like a nestling dove,
> From heav'n's dun concave shot a golden ray...
> Thou badst the softly kindling flame
> Pervade this peopled frame
> And smiles, with blushes ting'd, the work approv'd.[76]

As a final triangulation point for this chapter, consider the views of Thomas James Mathias, contributor to the *Anti-Jacobin*, Fellow of the Royal Society and staunch defender of the establishment. His vituperative poem

The Pursuits of Literature (1794) was published serially, and the fourth edition appeared in 1798, the same year as 'The Loves of the Triangles'. From Mathias's perspective, Jones lies at the apex of the triangle, while Darwin and Knight are relegated to baseness. Bracketing their style as evidence of 'the decline of simplicity and true taste in this country [to] filmy, gawzy, gossamery lines', he poured vitriol on Darwin's erotic botany and Knight's revelations about phallic worship.[77] In contrast, Mathias regarded Jones as a stalwart Christian who 'saw THE STAR, and worshipp'd in the East'.[78]

Despite their similarities, within a few years these three men were regarded in very different ways. Jones's reputation soared, Knight was remembered for his pronouncements on picturesque landscapes, while Darwin was condemned to near-oblivion until resuscitated through association with his grandson.

THE LOVES
OF THE PLANTS

BOTANY, WOMEN, AND
MORALITY

I love to feel events overlapping each other, crawling over one another
like wet crabs in a basket.

Lawrence Durrell, *Balthazar* (1958)

When the British Museum opened in 1759, its first director
was a sour, reclusive man called Gowin Knight (no relation
to Richard Payne). Being the sort of curator who regarded
the collections as his private property, Knight blocked off the corridor to
the lavatory so that he would not be disturbed by visitors walking past his
study, and complained about elderly readers nodding off in the overheated
library. Over two centuries later, the Museum still housed the British
Library, and it was still occupied by dozing researchers (of all ages). Sometimes I found myself relegated to a special table placed right beneath the
supposedly watchful eyes of staff who were as immersed in *Women's Own*
as I was in the eighteenth century. I always felt uncomfortable there, aware
that everybody else was aware I was reading classified material—which for
the British Library meant not state secrets, but pornography, including
Richard Payne Knight's *Priapus*.

I discovered that Darwin was not alone in mixing science and sex. The
Library's censored list included *Teague-Root Display'd*, ostensibly horti-
cultural but betrayed by its author's pseudonym of Paddy Strongcock.
Whoever Paddy S really was, he was well-informed about recent electrical experiments at the Royal Society, and expected his readers to appreciate how glass tubes become full of fire when they are stroked. However,
the Library vigilantes missed a similar piece by an extremely eminent
author—Samuel Johnson. He too was up to date with events at the
Royal Society, and his salacious descriptions of magnetic steel bars (you

rub them and they get stronger) show a detailed knowledge of the latest research.[1]

Discussing such texts can be difficult, but to ignore their existence is to distort the past, because they form a significant sub-genre of scientific popularization. One particularly relevant example is John Armstrong's *The Oeconomy of Love* (1736), a sex manual disguised as a poem (masturbation features prominently). Published in cheaper and cheaper editions for almost eighty years, at one stage it appeared bound together with a poem for women by William Hayley, one of Darwin's friends. Like Darwin, Armstrong was worried that his professional medical career would be adversely affected by writing in verse.[2]

Many activities that now seem innocuous used to carry sexual implications. Take gardening. During the eighteenth century, gardens belonged to country mansions, not to suburban semi-detacheds. Adorned with groves, temples, and grottoes, they were designed not only to advertise wealth but also to provide convenient backdrops for seduction. Men owned not only their grounds but also their womenfolk—Capability Brown was called 'the second monarch of landscape' as well as 'Lady Nature's second husband'. In Thomas Gainsborough's famous picture of *Mr and Mrs Andrews* (1748–9) at London's National Gallery, she sits passively in front of a tree as though she too were a plant rooted in the scenery, while he is holding a suggestively placed gun, clearly the possessor of her, his lands and his hunting dog.[3]

Gardening was an ideal occupation for leisured, wealthy gentlemen with a literary or artistic bent, and many fine (and not so fine) Augustan poems convey this metaphorical interplay of landscapes and female bodies, of planting and dressing, of beauty and seductiveness. Husbands and lovers lavished enormous sums of money on decorating their women with expensive clothes and jewels. Similarly, they poured cash into redesigning the landscape, expecting in both cases to be gratified with delightful views and titillating entertainment. When describing the delights offered by their grounds, male owners used loaded terms such as 'Secret Satisfaction', 'Recesses' and 'Penetration'. Stowe, one of the eighteenth century's most admired and copied gardens, still survives, although now a tourist attraction attached to an expensive school. The place names on its map—the

Lady's Temple, the Temple of Venus, the Queen's Valley—reveal how its landscape was intended to be both feminine and enticing.

Country estates might not have appeared artificial, but in reality a great deal of hard work and money went into constraining nature and making her behave as desired. Jagged profiles were smoothed into curves, unruly brooks were rechannelled into serpentine streams, colourful flowers were confined in beds by formal gravel paths. Similarly, the vagaries of the English language were corseted into heroic couplets, as in Alexander Pope's instructions for creating a deceptively demure Dame Nature:

> But treat the Goddess like a modest fair,
> Nor over-dress, not leave her wholly bare;
> Let not each beauty ev'ry where be spy'd,
> Where half the skill is decently to hide.[4]

Darwin first became interested in flowers after falling in love with a dedicated gardener, Elizabeth Pole, the colonel's wife from Derbyshire. Before long, Darwin had bought a plot of land about a mile away from his Lichfield home, a steep narrow valley referred to by his family as 'the botanical garden', which later became his working title for *The Loves of the Plants*. Enlisting the help of his small son Robert, Darwin dammed and diverted the brook, got rid of the weeds and stones, and arranged some carefully chosen trees and flowers to create an ornamental garden with lakes, shrubberies, and waterfalls.[5] Sexual vocabulary often drew on aqueous images—overflowing passions, torrents of desire, the sluice gate of prostitution. Darwin featured his garden in *The Economy of Vegetation*, presenting it 'as adapted to love-scenes' with its 'grotto surrounded by projecting rocks, from the edges of which trickles a perpetual shower of water'.[6]

Darwin tamed his rough plot into a metonym for his female object of desire—a well-tended garden that he adorned expensively and with meticulous attention. Robert and his young brothers may or may not have appreciated the thoughtfully constructed sexual symbolism, but they presumably never saw the ardent love poem Darwin sent his 'fair Eliza' to thank her for a gift of some flower-embroidered cuffs (which, according to him, he smothered with kisses each morning). The second stanza not only suggests the preoccupations of eighteenth-century men when they were designing landscapes, but also shows that women were expected to appreciate such geographical imagery too:

> If brighter scenes from other's hills extend,
> If others' vales more fragrant flowers defend,
> Each passing swain, with rural charms inflamed,
> Eyes the gay scene, and drinks the gale unblamed.[7]

Making gardens seductive happened not only at the large-scale level of landscapes, but also in the micro-detail of flower descriptions.[8] Some plants symbolized human reproductive organs, and nudge-nudge jokes appeared in gossip columns: '*Cuckeldom—A corniform* plant which grows in several *beds* has been found to flourish successfully in *shrubberies*.'[9] Some flowers carried suggestive names such as Venus's Navel-wort and Venus's Lookingglass, but during the second half of the eighteenth century, plants acquired technical Latin labels (for these, *omphalodes linifolia* and *triodanis perfoliata*). Although many people continued to use the more evocative vernacular names, this new language of flowers enabled botanists to exchange information and specimens internationally. The system was introduced by the Swedish botanist Carl Linnaeus, now celebrated as the great Enlightenment taxonomist who imposed order on the unruly natural world. He proposed that every plant should carry its own unique two-part identification. Lemon trees, for instance, would be called *citrus limon* to distinguish them from their close relatives, orange trees, or *citrus aurantium*.

Linnaeus' system now seems so familiar that it feels intuitively right, but it is based on arbitrary criteria and had many critics. There could be no single correct method, and other botanists classified plants by the colour of their flowers or the shape of their leaves. Linnaeus chose to focus on reproductive organs, even though many plants are hermaphrodites, which carry both male and female parts. First he set up twenty-four classes by counting the number of male stamens in the flower; he then subdivided each class into less important orders based on the number of female pistils.

As his model for this system, Linnaeus turned to human relationships. The prejudices of Enlightenment Christian moralists are built into this scientific plan for plants, which Linnaeus sketched out using anthropomorphic terms such as 'bride' and 'marriage'. In his scheme, the most basic division is between male and female, and Linnaeus gave priority to masculine characteristics. In other words, he imposed onto the plant kingdom the sexual discrimination that prevailed in the human world. His first level of ordering depends on the number of male stamens: only the sub-groups are determined by the number of female pistils.

From the dominant position enjoyed by Linnaeus and his male contemporaries, this botanical scheme carried a great advantage: it made his subjective organization of plants appear as though it were natural, even God-given. Linnaeus had mapped human society onto the botanical world, but from then on men of science could argue in reverse. Since sexual hierarchies prevail in nature, male supremacy must also—so the distorted logic runs—be appropriate for people; this argument conveniently forgets how the sexual ordering was inferred from society in the first place. Through this closed loop, Linnaean classification not only mirrored social prejudice, but reinforced it.

This man who introduced eroticism into botany was a home-loving reactionary who refused to let his daughters learn French in case they lost their appetite for housekeeping. He equated sexuality with marriage rather than with promiscuity, and regarded women as wives and caregivers, not individuals with their own desires and ambitions. There was a problem, however: many orders of plants had unequal numbers of stamens and pistils, and so could not possibly correspond to conventional relationships. To describe these unorthodox arrangements, Linnaeus used expressions such as 'concubine' and 'clandestine marriage'.

In Britain, the overriding objection to Linnaeus' scheme was that it put sex right at the heart of botany, thus making it unsuitable for women. Writing in Latin to reach an international audience of scholars, Linnaeus clearly spelled out the analogies between the reproductive organs of flowers and people. 'The calyx is the bedchamber,' he explained in 1735, 'the filaments the spermatic vessels, the anthers the testes, the pollen the sperm, the stigma the vulva, the style the vagina.' Such explicit explanations seemed scandalous—'too smutty for British ears', one critic spluttered. And especially so for one half of the population: one clergyman protested, 'Linnean botany is enough to shock female modesty.'[10]

In some ways, studying flowers seemed an ideal pastime for women—not too taxing mentally, a gentle occupation that could be carried out peacefully at home but also involved some therapeutic exercise. On the other hand, botanic vocabulary vibrated with sexual innuendo. Darwin's chief educational rival was a fellow member of the Lunar Society, William Withering, a physician famous for curing heart problems with medicines made from foxglove (*digitalis*), and a best-selling author on plants. To sanitize Linnaean botany, Withering translated contentious words into harmless but meaningless English equivalents such as 'chives' and 'pointals'. Explicitly writing for women, he aimed to make botany 'as healthful as it is innocent' so that it 'leads to pleasing reflections on the beauty, wisdom, and the power of the great CREATOR'. When equipped with Withering's bowdlerized botany, from which both sex and Latin had been expurgated, women could discuss flowers safely without being accused either of sexual impropriety or of pedantry.[11]

Darwin thought Withering was prudish. In his opinion, 'Linnaeus might certainly be translated into English without losing his sexual terms...and yet avoiding any indecent idea'. Drawing up a tentative vocabulary list, Darwin rejected 'viragoes' and 'cuckoldoms', but decided that 'polygamies' and 'male-coquetts' were acceptable.[12] To consolidate Linnaeus' status, Darwin set about producing an English version of *Systema Vegetabilium*, Linnaeus' large technical book explaining how to classify all the members of the plant world—trees and flowers as well as vegetables.

To help him out, Darwin recruited some volunteers and optimistically set up the Botanical Society of Lichfield. Even at its peak, this grand-sounding institution only had two other members, neither of whom contributed much. One was Brooke Boothby, a local baronet's son who had recently settled down in Lichfield after an extended and expensive Grand Tour. A great admirer of the French philosopher and social reformer Jean-Jacques Rousseau, Boothby is the elegant subject of an unusual portrait by Joseph Wright of Derby, who painted him reclining in a grove, a volume of Rousseau in his hand. Although Darwin was a less enthusiastic follower, he had discussed botany with Rousseau during his stay in England, and they allegedly corresponded for several years.

Darwin found it hard to enlist other friends, but he did manage to involve a local man called William Jackson, whom sharp-tongued Anna Seward regarded as an unscrupulous social climber, 'a turgid and solemn coxcomb' who had 'sprung from the lowest possible origin'.[13] William Jackson is a common name, but when the library catalogue produced one from Lichfield Close, I guessed he might be Darwin's ally. Jackson's *Beauties of Nature* (1769) turned out to be a tedious, conventional tract proclaiming God's munificence in creating such a splendid world. I was rewarded for ploughing through homilies about the virtues of temperance by discovering some crude poems tacked on to the end. Whoever Jackson was, his erotic botany makes Darwin seem subtle:

> Oh! Change her Jove, into a Rose,
> Myself into a Bee;
> That I for ever hidden close,
> May in her petals be.
> There will I build my nectar'd Home,
> My curious Dome contrive;
> Her Lips shall be my Honey-comb,
> Her folding Leaves my Hive.[14]

Because of Darwin's many other commitments, the new Society's translation project was put on hold, and might never have got off the ground at all if Elizabeth Pole had remained unavailable. At the end of 1780 her husband died suddenly, and within a few months Darwin had abandoned his Lichfield botanic garden and moved into her house near Derby. Freshly wed, Darwin was delighted with his bride. She was, he wrote to his Lunar friend Thomas Day, 'possessed of much inoffensive vivacity [and] like myself, she loves the country and retirement...'[15]

Even so, rural seclusion proved trying. While looking for a city house to use as a surgery, Darwin needed something to keep him occupied. His first step was to translate a few pages of Linnaeus' *Systema Vegetabilium*, and send them out for comment to forty leading botanists. Darwin was adept at soliciting patronage: he once advised Josiah Wedgwood not to betray his artisan origins by using overly obsequious language. In order to secure his most prestigious British backer, Joseph Banks, Darwin sent off a carefully judged letter asking permission to dedicate his translation to him.

Banks operated an international correspondence network from his base in London's Soho Square. Twenty thousand letters to and from Banks still survive, stored in archives scattered around the globe. Experts estimate that he wrote forty thousand himself, which averages out at about three a day—some of them very long—right through his forty-two-year regime as president of the Royal Society. Banks was extremely generous to Darwin, offering him advice and lending him books from his own private library. In return, when Darwin's translation finally appeared, he included a tribute to Banks, praising him in appropriately lavish eighteenth-century terms.

A thousand pages long, *A System of Vegetables* (1783) was a serious work of scholarship. Botanical insiders knew the true identity of the Lichfield Botanical Society credited on the title page, and they praised Darwin's industry in producing these two weighty tomes. Determined to convey the nuances of Linnaeus' Latin labels, Darwin consulted a fellow Lichfield citizen, the English lexicographer Samuel Johnson. Sometimes Darwin's concern for accuracy became obsessive: can his correspondents really have been interested in considering the niceties of whether 'eggshape' should be changed to 'eggshaped'? On the other hand, this was a time when technical terms were being introduced into botany—even 'petal' was only coined in the late seventeenth century. 'Floret', 'bract' and 'leaflet' are now in common use, but Darwin was one of their earliest users.

At the same time as carrying out this conventional type of translation work, Darwin was also wondering how he could convey Linnaeus to non-experts, to ordinary men and women who wanted to learn more about the new classification scheme. Although he did not realize it, he was dreaming up his first poetic best-seller—*The Loves of the Plants*.

CHAPTER FOUR

The Loves of the Plants

Pink petals flare in the hedge:
Rosa canina, rose of the dog days.
Gently I splay the petals, bend and sniff.
The whole flower gives off sweetness,
Pungency deeper in.
Such heat this evening, come away from the path.

Clive Wilmer, 'Dog Rose in June' (2006)

In its electronic catalogue, the Cambridge University Library has forty-nine entries for works by Erasmus Darwin.[16] Arranged alphabetically, they start with *Amori delle Piante*, an 1805 Italian translation of *Loves of the Plants*. It was some time before I worked down the list to item number 38, which has the intriguing title of *Pussey Cats' Love Letters*. On asking for it to be retrieved from one of the locked rooms scattered around the Library's six floors, I was astonished to be handed a small envelope containing a book that measured only 5 x 7 centimetres. Despite its tiny size, it was easy to read and lovingly bound in leather with gilt lettering—an Alice in Wonderland version of the eighteenth-century tomes I was used to holding.

Even the librarian had never come across a book like that before. It was, I discovered, an example of microbibliomania (from Greek: 'small-book madness'). The craze started in Great Britain and spread in the early twentieth century to the United States, where the most famous microbibliophile was President Franklin D. Roosevelt. The example lying in the palm of my hand was printed in Collingswood, New Jersey, in 1934; it came from the private press of William Lewis Washburn, and a handwritten note in red ink told me it was the thirty-ninth of only sixty-four copies. Washburn was a commercial printer, but for over ten years he spent his evenings down in the basement hand-crafting twenty-two sets of miniature volumes.[17]

The contents seemed as whimsical as the format—a playful exchange of letters between Darwin's Persian cat Mr Snow and Anna Seward's Miss Po Felina. This is the opening of Mr Snow's epistle:

> Dear Miss Pussey
>
> As I sat, the other day, basking myself in the Dean's Walk, I saw you...washing your beautiful round face, and elegantly brindled ears, with your velvet paws and whisking about, with graceful sinuosity, your meandering tail. That treacherous hedgehog, Cupid, concealed himself behind your tabby beauties, and darting one of his too well aimed quills, pierced, O cruel imp! my fluttering heart.[18]

Signed 'your true admirer Snow Grimalkin', this letter continued as a playful proposal of marriage. The next day, Miss Po Felina composed her reply. 'I am but too sensible of the charms of Mr Snow,' she began, clearly paving the way for a major reservation. It soon followed. 'I admire the spotless whiteness of his ermine, and the tiger-strength of his commanding form,' she wrote, yet worry lest he 'retain the extreme of that fierceness, too justly imputed to the Grimalkin race...Marry you, Mr Snow, I am afraid I cannot.'[19]

I later discovered that although this clandestine correspondence was exchanged in 1780, Seward published it only in 1803, the year after Darwin's death. Were her motives for revealing it as disinterested as she claimed? According to contemporary visitors, Lichfield was dominated by widows and spinsters, and so Seward's rejection of an offer to improve her social status through marriage seems unusual.[20] Perhaps Seward was playing too cool a game against Elizabeth Pole, her rival for Darwin's attentions. She may have reasoned that the widow had already had enough experience of one bad-tempered husband twice her age, and so however ardent Darwin's courtship might be, was hardly likely to prefer this pockmarked, overweight, and elderly suitor to his younger competitors. Only a few months later, Pole consented to marry Darwin on condition that he move in to her house near Derby. Seward may have felt that when Darwin married Pole, things hadn't turned out quite as she planned.

My suspicions about Seward's detachment were strengthened when I read more of her *Memoirs*. Apparently she too had been invited to visit his botanic garden—and while there, she had written a poem about plants that he incorporated into his own work without crediting her. Darwin's popular poem was therefore inspired by Seward, and it was originally planned as a joint effort, consisting of her verses amplified by his scholarly notes. At least, that was her version of what happened.

Was it just discretion that made her wait until the year after Darwin's death before publicly accusing him of plagiarism? Why was her longest work of literary criticism directed against Darwin's *Botanic Garden*? Did Darwin include in his own poems lines that Seward had written? Biographers of Seward and Darwin are still battling it out.

First Impressions

The first time I read *The Loves of the Plants* I was tempted to laugh, but I was too bored. I thought those cavorting floral goddesses were ridiculous, and I found the footnotes easier to understand than the verses. I realized that I must be missing something. Reviewers had puffed Darwin as a new Pope or Milton: even the hard-headed industrial chemist James Keir gushed that in this 'exquisite poem...the graces themselves seem to decorate the temple of science with their choicest wreaths and sweetest blossoms'.[21] Darwin's personal measure of success was the number of copies sold, and he boasted to his Lunar friends about having created a commercial product as profitable as the goods being manufactured in their Midlands factories.

Why, I wondered, had so many people wanted to buy this poem supposedly written by the Botanical Society of Lichfield? Even on a casual inspection, it looks completely different from a modern book of poetry. Many of its pages are dominated by learned footnotes, while three prose interludes (mostly ignored by modern commentators) separate the verses into four sections called cantos. This musical label originated when poetry was designed to be sung, so that minstrels could pause as if between book chapters, and many poets divided their works in this way.

Although this set of around 1,700 clunky lines will never become my favourite bedtime reading, familiarity has bred not contempt but greater respect for the poem and its author. So much that initially seemed bizarre to me would, I now realize, have been instantly recognizable to Darwin's fans. Stories that used to be as well-loved as Cinderella or Snow White are now restricted to experts. For almost a millennium, all educated people were drilled from early childhood in two great traditions: classical literature and the Bible. During the twentieth century, within the lifetime of many people alive today, this cultural heritage vanished at an astonishing speed, so that Achilles and Abraham, Jove and Joseph, Mars and Moses are now becoming jumbled up with Santa Claus into the same fairy-tale category as Hansel and Gretel.

Stretching that gulf between the centuries still further, the underlying rationale for telling stories has changed. Nowadays, readers expect authors to invent new plots, characters, and emotional dilemmas. In contrast, poets such as Chaucer or Milton started with a familiar tale—often a classical myth or an episode from the Bible—and told it in a fresh way, perhaps picking a different person as the main narrator, or imagining what happened later. The idea was not to summon up an unprecedented situation, but instead to shake up expectations by thinking about an old drama from an unusual angle.

Another obstacle to understanding the past is the casual assumption that because poems belong to literature, they should be about feelings and imagination, not about facts and knowledge. But just like novelists or historians, poets use words to articulate arguments and opinions as well as to elicit emotions. In 1959 the Cambridge scientist C. P. Snow consolidated the Two Cultures divide, divorcing the arts and the sciences so effectively that his biased perception has distorted the historical landscape ever since. If only he had taken more notice of his own warning. 'Attempts to divide anything into two,' Snow counselled in his seminal lecture, 'ought to be regarded with much suspicion.'[22]

During the eighteenth century, science and poetry were definitely not separate domains. Poetry was entertaining as well as educational, emotive as well as edifying—and it sold well. Many religious poets celebrated the glory of God by describing His creation in scientific terms, but the most famous Enlightenment counter-example to Snow's claim is James Thomson, who taught physics in London before turning to writing. His long poem *The Seasons*, repeatedly republished, emulated, and translated, juxtaposes verse accounts of Newtonian optics with lyrical descriptions of natural phenomena. Reading Thomson might be a more pleasurable experience than reading Darwin, but they both used the genre of poetry to express ideas about nature, religion, and contemporary science.

While learning about *The Loves of the Plants*, I had already encountered selected stanzas, but I decided that the time had come to read it from cover to cover. Sometimes when you pick up a book armed in advance with other people's impressions, you find yourself wondering whether they've even been looking at the same text (and why do so many academics seem to lack any sense of humour?). What follows is what I personally experienced. I've aimed to make my remarks helpful, but I also hope that you will want to argue against my particular judgements, summaries, and choice of extracts. If so, please adopt my strategy of going back to Darwin's original so that you can form your own opinion. By far the best way to think about a work of literature is to study it for yourself.

Beginnings

In the pre-electronic era, reading *The Loves of the Plants* systematically would have entailed sitting quietly on a hard chair in the Rare Books room of a library, surrounded by intellectual sleuths, those academic Dickensian clerks engrossed in copying out long tracts in pencil (pens are forbidden). It's gratifying to feel part of a privileged coterie, especially if you're looking at the very same copy once held by somebody famous. But although there's no substitute for handling an old leather-bound text, there's also no substitute for sitting in a comfy chair, drinking a cup of coffee, and feeling free to highlight potential quotations. Darwin's poetry is now available online and as cheap paperbacks, but—and this is a big BUT—not even the contents of these modern versions are identical to the originals, let alone their appearance.

For *The Loves of the Plants*, the differences start right at the beginning, with the frontispiece. The reprint (not a facsimile) has no pictures at all, while the online version I consulted reproduced only one, and that was from a later edition. In contrast, in 1789, when Darwin's first readers opened up their new purchase, they could browse through a handsome leather-bound tome printed with wide margins on heavy paper. At the beginning, they would have found a black-and-white picture of 'FLORA at Play with CUPID' (Fig. 2). This was drawn by one of Wedgwood's pottery designers, Emma Crewe; for later editions, Darwin attracted a more famous artist—Henry Fuseli—who showed the goddess of flowers admiring herself in a mirror as four beautiful young women help her to dress.

Darwin had met Fuseli in London while on a honeymoon trip with his second wife, and shortly afterwards, he contributed a verse to boost sales of what is now Fuseli's most famous image, *The Nightmare*, in which an incubus squats on the chest of a voluptuous sleeping woman while a horse peers voyeuristically around the edge of a curtain (although etymologically, the word 'nightmare' has nothing to do with a female horse). Economically, Darwin recycled and also amplified these lines by incorporating them in *The Loves of the Plants*, whose later editions include illustrations by Fuseli.[23]

'Frontispiece' was originally an architectural term meaning the decoration surrounding the front door—an invitation to enter the building. Darwin's frontispiece enticed his readers to buy his book by condensing its theme into a single allegorical picture. In Crewe's engraving, Cupid is marching off with some gardening tools, while the goddess Flora holds his bow and arrows. Bedecked with flowers, armed with the weapons of love, she lounges languidly, gazing out of her bower as if searching for a suitable heart to target amongst her spectators. Or as one critic put it, the artist 'overstepped the modesty of nature, by giving the portrait an air of voluptuousness too luxuriously melting'.[24] Facing this seductive image, the formal title page carries a Latin quotation: taken from a wedding poem, it describes trees bending together and whispering love.[25]

Determined to present his poem as a serious work of scholarship, Darwin provided copious notes and a technical summary of Linnaeus' classification system. He also illustrated his poem with botanical drawings,

Figure 2 Emma Crewe, *Flora at Play with Cupid.* The original frontispiece of *The Loves of the Plants* (1789). © Trustees of the British Museum.

which are missing from the paperback and are misleadingly coloured in the online version: at the end of the eighteenth century this type of printing was exclusively in black and white, although illustrations might subsequently be hand-coloured by purchasers or low-paid piece-workers, usually women.

Darwin's opening 'Advertisement' is also missing from the paperback copy. All these details about editions may seem pedantic overkill, but to understand the aims of the *Anti-Jacobin* parodists, I wanted to encounter Darwin as they had done. For them, Darwin's 'Advertisement' was not just an extra, but an integral part of his poem's design. Whereas their parodic version started with a ponderous prose introduction supposedly by William Godwin, in his own 'Advertisement' Darwin tried to justify the frivolity of his verses by announcing that he would 'inlist Imagination under the banner of Science'. His most famous dictum, this rapidly became a catchphrase, a quasi-advertising slogan for his work. In the satirists' rendition, supposedly by Godwin, it reads: 'I am persuaded that there is no Science, however abstruse, nay, no Trade or Manufacture, which may not be taught by a Didactic Poem. In that before you, an attempt is made (not unsuccessfully I hope) to *enlist the Imagination under the banners of Geometry. Botany* I found done to my hands.'[26]

Next in the original *Loves of the Plants* comes a short PROEM (a term introduced into English by Chaucer—it means preamble), which is addressed to the 'GENTLE READER!' and decorated with a podgy cherub holding a butterfly. Although only four sentences long, it baffled me. This is how it starts: 'Lo, here a CAMERA OBSCURA is presented to thy view, in which are lights and shades dancing on a whited canvas, and magnified into apparent life!'[27] The Latin 'camera obscura' means 'darkened room'. Imagine standing inside a closed space with a single small hole on one side, through which the sun is shining; if you look at the opposite wall, you will see upside-down images cast on the surface by the sunlight. Miniaturized boxed versions were used by anatomists to make accurate drawings of skeletons, or by mediocre artists to sketch a landscape accurately: they could trace the patterns of light directly onto a piece of paper acting as a projection screen.

In this pre-cinema era, the camera obscura also offered a brand-new type of entertainment. Alexander Pope described the visual effects he created in the grotto of his Thames-side garden at Twickenham: 'When you shut the Doors of this Grotto, it becomes on the instant, from a Luminous Room, a *Camera obscura*; on the Walls of which all objects of the Rover, Hill, Woods, and Boats, are forming a moving Picture in their visible

Radiations.'[28] Like him, fastidious landowners often spent years digging out subterranean vaults, lining them with complex patterns of minerals, or even installing steam-driven cascades. Although Pope's grotto no longer exists, I have visited several others that survive from the eighteenth century, including the poet John Scott's shell-lined fantasy in Ware, a series of chambers stretching 20 metres underground into the chalk hillside. Rescued from the bulldozers by a last-minute preservation order and crammed up against a modern housing estate, it was once 'a fairy hall' that attracted tourists from London. Scott's puritanical brother strongly disapproved: as Richard Payne Knight and other gardeners knew well, such dark enclosed chambers resonated with sexual metaphor.[29] Darwin's verbal enticement signals that erotic delights lie ahead.

At the same time that Darwin was writing *The Loves of the Plants*, the philosopher and legal reformer Jeremy Bentham was designing his panopticon (from the Greek for 'all' and 'observe'). In this high-surveillance building, a central guard could monitor prisoners under his charge without himself being seen, as if he were an invisible omniscience. Similarly, by imposing order on an unruly collection of plants, Linnaeus resembled a Divine Architect whose hidden laws are undetectable to lesser beings. As property owners, gentlemen could look down on their estates from above, whence they could perceive a logical arrangement invisible to those toiling in the flower beds. In contrast, when Darwin invited his guests to accompany him around his imaginary garden, he was offering a ground-level experience with which women and labourers were familiar.

The Proem's second sentence mentioned 'a great Necromancer' of whom I'd never heard—P. Ovidus Naso. It turned out that I had: on consultation, my poet told me (rather condescendingly) that this is the full name of the Roman poet Ovid, who lived around the time of Christ and was extremely popular in the eighteenth century. His most influential work was *Metamorphoses*, which, as its name suggests, is about transformations. Human beings are changed into trees or stars, animals into stone or flowers, and vice versa in various combinations. Conceptually, such continuous alteration is at odds with the traditional Christian view of a cosmos fixed by God at the Creation, but it fits well with evolutionary views of a progressive universe. In particular, Ovid opens his collection of stories by describing how an original formless mass, or Chaos, was transformed into an organized universe, a theme that became central to Darwin's later work.

Reading about Ovid made me think again about Crewe's frontispiece (see Fig. 2). In pre-Christian Italy, Flora was a fertility goddess, and the annual springtime festival included striptease shows by prostitutes. By emphasizing Flora's sexuality, these rituals consolidated the division of Roman women into two stereotypes—the sober matron or housewife, and the harlot unable to control her desires. While Ovid was composing the *Metamorphoses*, he was also working on another long poem called the *Fasti*, a versified calendar of the Roman year (*dies fasti* were the days on which it was legal to transact business), in which he glorified Flora as a powerful goddess of flowers. In eighteenth-century England, she still carried mixed messages of divinity and eroticism (although in neither Darwin nor Ovid is there any suggestion that she might have a mind as well as a body).[30]

Darwin knew Ovid's poems intimately himself, and assumed—as did the *Anti-Jacobin* satirists—that readers would too. However recherché a classical reference might be now, two centuries ago it needed no spelling out. For example, a caricature by James Gillray (of 1795) parodies Joseph Banks as a giant butterfly spreading his wings beneath a crowned sun and rising up 'from among the Weeds & Mud on the Banks of the South Sea'. Banks was, Gillray mocked, like a caterpillar being bred from the mud by the sun of royal patronage—an image of transformation that originated in Ovid but was borrowed by Shakespeare for *Antony and Cleopatra*.[31] Even women would be familiar with the plots and characters, although the strict grammatical rules of Greek and Latin were deemed too rational for girls to tackle: while small boys struggled to learn these languages from printed primers, their sisters were taught French and Italian through conversation. But Darwin was determined not to deter any of his female readers by seeming too scholarly, and he reassured them that something light and frothy lay in front of them—his poem would, he promised, lead them on a stroll around an enchanted garden.

Darwin's small plot in Lichfield was a scaled-down version of those grandiose estates designed according to the precepts of Capability Brown or Richard Payne Knight and now favoured by the National Trust. The more famous ones attracted high-class tourists for picnics, sightseeing, seduction and—what might seem stranger now—moral edification. Land-scapes were intended not only to provide striking views and romantic set-tings for courtship, but also to inform and to educate. Most famous of all, Lord Cobham's garden at Stowe had an international reputation. The tourist brochure was a long poem that conducted armchair visitors around the estate, entertaining them in their imaginations with dramatic per-formances featuring mythological characters. Similarly, as you read through Darwin's poem, you wander in virtual reality through intercon-nected glades and grottoes watching a continuous performance unfold in front of you, as flowers attuned to the hours of the day turn their heads or open and close their petals.[32]

It's taken me several pages and many hours in libraries to unravel what Darwin expressed in two short paragraphs. At the time, perhaps many readers skipped over this preamble, and turned straight to the poem itself. That is what I did next.

The Four Cantos

These are some typical lines from the first canto (minus the far longer paragraphs of notes accompanying them):

> When Heaven's high vault condensing clouds deform,
> Fair AMARYLLIS flies the incumbent storm,
> Seeks with unsteady step the shelter'd vale,
> And turns her blushing beauties from the gale.—
> *Six* rival youths, with soft concern impress'd,
> Calm all her fears, and charm her cares to rest.—[33]

Turning to the foot of the page for help, I learnt that amaryllis flowers (six stamens, one pistil) close their petals and turn down their bells for protec-tion from rain and cold—or as Darwin's male chauvinistic botany would have it, the frightened female is soothed by six solicitous suitors.

Like many of Darwin's notes, his information about amaryllis mixes together technical terms—fecundation, pendant attitude, lacerated corol—with homely analogies, such as sheltering under an umbrella or turning like a weathercock. For the *Anti-Jacobin* satirists, Darwin's pompous style and pernickety detail provided perfect targets. Sometimes they parody him by stating the obvious: '*Pulley*—So called from our Saxon word, to PULL, signifying to pull or draw' or '*Circles*—*See* Chambers's Dictionary—Article Circle'. Others offer sexual allusions: '*Hydrostatics*—Water has been supposed, by several of our philosophers, to be capable of the passion of Love.—Some later experiments appear to favour this idea—Water, when pressed by a moderate degree of heat, has been observed to *simper*, or *simmer* (as it is more usually called).' Although I did mange to pick up a few Latin puns (for example, understanding the joke in 'a pebbly channel, inlaid with *Differential Calculi*' depends on knowing that calculus is a small stone as well as a kind of mathematics), there must remain many jokes that I have failed to detect.[34]

Returning to Darwin, I read on beyond 'Amaryllis', bemused (but certainly not gripped) by clandestine marriages of seaweed and liverwort, by the ambrosial lives of grasses, by the spurned dusky bride of cypress. One reader was horrified at Darwin's report that the cypress excludes his wife from his bed, and begged the author to desist from libelling 'harmless trees and flowers'.[35] The poet was clearly enjoying himself, hovering on the verge of self-mockery by using such a formal classical apparatus for a non-heroic topic. The notes provided me with enough factoids for a year of cocktail parties: a downy fern from China was thought to be a lamb; twice-weekly baths slow down the process of ageing; the same Greek word means both fig and cuckoo, because they arrive at the same time in the spring. I also learnt how some eighteenth-century gentlemen spent their time—feeding piglets with madder on alternate weeks to show that their bones become striped with red.

All very fascinating, but I was still perplexed. Why should the *Anti-Jacobin* be so concerned about Darwin's poem? When I embarked on *The Loves of the Plants*, I was expecting to find little but sexual innuendo. However, as I persevered further, I began to wonder whether most readers had given up after a few pages. In the second canto, this supposedly frothy piece of botanical soft-porn transmutes into a celebration of the industrial processes being introduced by Darwin's Lunar Society colleagues in their Midlands factories. For instance, flax—which has five stamens and five pistils—is used for weaving linen, and Darwin combines romantic and mechanical images of spinning:

> *Five* Sister-nymphs with dewy fingers twine
> The beamy flax, and stretch the fibre-line;
> Quick eddying threads from rapid spindles reel,
> Or whirl with beaten foot the dizzy wheel.
> —Charm'd round the busy Fair *five* shepherds press,
> Praise the nice texture of their snowy dress,
> Admire the Artists, and the art approve,
> And tell with honey'd words the tale of love.[36]

A couple of pages later, Darwin is on to cotton, and so excited about a recent invention, Richard Arkwright's spinning machine, that he temporarily forgets about sex:

> With quicken'd pace *successive rollers* move,
> And these retain, and those extend the *rove*;
> Then fly the spoles, the rapid axles glow,
> And slowly circumvolves the labouring wheel below.[37]

In his notes, Darwin emphasizes that industrialization brings progress—Arkwright's machine is, he claims, not only much faster but also performs '*better* than can be done by hand'.[38] I imagined the *Anti-Jacobin* satirists bristling at that sentiment, and I was not surprised to find these particular lines singled out for attack. In their version, 'fair Trochais' (identified in a footnote as 'The Nymph of the Wheel') has been peering down from the chimney at 'Her much loved *Smoke-Jack*' whose 'various parts…in one object end' (if Linnaean botany can be pornographic, then so too can political satire):

> The spiral *grooves* in smooth meanders flow,
> Drags the long *chain*, the polish'd axles glow,
> While slowly circumvolves the piece of beef below:
> The conscious fire with bickering radiance burns,
> Eyes the rich joint, and roasts it as it turns.[39]

Throughout this second canto, Darwin joyfully (if sometimes at dreary length) celebrates human achievement—especially English achievement. For the third, he switches into a dark melancholic mood. Instead of emphasizing the medical or industrial applications of plants, Darwin focuses on their deleterious effects, featuring poisons, hallucinogens, and grapes. (In his eyes, alcohol was the major cause of chronic illness, and he advised patients to adopt the Muslim habit of abstention.) Death, disease, and disaster appear on every page, climaxing with a melodramatic protest against the evil of slavery:

> E'en now, e'en now, on yonder Western shores
> Weeps pale Despair, and writhing Anguish roars:
> E'en now in Afric's groves with hideous yell
> Fierce SLAVERY stalks, and slips the dogs of hell;[40]

Feeling that I was beginning to understand why the *Anti-Jacobin* should choose this poem to satirize, I moved on to the fourth and final canto, when the sun is setting but 'tardy Spring' is arriving. My new-found confidence in my powers of comprehension was short lived, as I struggled through swans and boars to the ending, a splendid multiple wedding on Tahiti:

> A *hundred* virgins join a *hundred* swains,
> And fond ADONIS leads the sprightly trains;[41]

For me, an Adonis complex meant over-exercising in the gym, but what had that got to do with flowers? Why did the note to that couplet describe hollow trees with living branches? And what was so special about Adonis that he formed the poem's climax?

Eventually, I fathomed out what must have seemed obvious to Darwin's readers. Banks had recently travelled to Tahiti, where British astronomers observed the Transit of Venus, measuring the planet's progress as it appeared to move across the sun. By imagining Venus's pleasure at a 'promiscuous marriage' between a hundred flowery grooms and brides, Darwin describes a nature who is fecund and polygamous, ruled by her own laws and not by those of a restrictive God.

Any eighteenth-century schoolboy would have picked up where Adonis fitted in. Lacking their mythological education, I turned to Ovid's *Metamorphoses*, where I read that Venus (aka Aphrodite) is in love with Adonis, who was born from inside a tree. Despite her warnings, he dares to hunt a wild boar that turns out to be a previous lover in disguise, and sinks his tusk into Adonis's groin (yes, there is much scholarly debate on what that might mean). Venus conveniently happens to be flying overhead in her swan-drawn chariot, and descends to hold the dying Adonis in her arms. Grief-stricken, she sprinkles his blood with nectar, causing each drop to be transformed into a red but short-lived anemone. Adonis goes down to the underworld, where Persephone is delighted to see him again (again? I realized I'd missed a bit... Venus had handed Adonis over to Persephone as a baby for safe keeping, but changed her mind when he grew up into a beautiful young man). To resolve the tussle for ownership between the rival goddesses, Zeus intervenes and says Adonis must spend half the year with each.[42]

Allegorically—and this is the important point for Darwin—Adonis and Persephone represent a continual cycle of death and rebirth. In autumn, nature is stained with the blood of dying Adonis, and plants die away to reappear in the spring. The poem ends as night falls over the garden, but Adonis's marriage indicates that the flowers will come to life again in the morning.

The more of *The Loves of the Plants* I read, the more complex it seemed. That made it difficult for me to understand, but also explained why it was so popular and why Darwin's critics took it so seriously. I decided to start again, but this time concentrating on the three prose interludes between the cantos.

Philosophical Prose

When Galileo Galilei wanted to convince his readers that the Earth moves, he wrote an imaginary conversation between his supporters and his opponents. Ostensibly a neutral debate, in reality it was a three-cornered defence of his own position, in which Simplicio (whose intellectual abilities matched his name) put forward the objections that were easiest to knock down: the Pope was not pleased to see his own point of view expressed by Simplicio.

In *The Loves of the Plants*, Darwin used the same literary device to justify his didactic poetry by creating fictional dialogues between himself—*Poet*—and an imaginary *Bookseller*. The word 'chutzpah' didn't enter the

English language until 1892, but Darwin certainly possessed it—in these prose sections, he corrects Shakespeare's punctuation, compares his own poetry with Homer's, and brackets himself with Newton as a scientific expert.

Modern scholars rarely refer to these three interludes, and I imagine many contemporary readers skipped over them.[43] But reviewers were certainly paying attention, if only near the beginning. The passage that annoyed them most appears in the first interlude, where Darwin dared to alter one of Pope's lines. In his imaginary dialogue, *Poet* maintains that prose is the medium for abstraction: poets should restrict themselves to what they can see. Pope was wrong, *Poet* claims, to write 'And Kennet swift for silver Eels *renown'd*', because renown is an invisible concept: a truly poetic version would be 'And Kennet swift, where silver Graylings *play*.'[44]

Abandoning both botany and sex, in these interludes Darwin assumes the role of literary critic and philosopher. According to him, a poet's job is to summon up a scene in the reader's imagination, and he includes physiological explanations to back up his insistence on the importance of visualization. Darwin was a poet of John Locke's Enlightenment, when to see is to know, to be illuminated is to understand, and the eyes are the most important route to the brain. The imagination resembles a physical faculty based in vision—the power to perceive what is not actually present but might be. The Romantics thought differently. For them, imagination is untrammelled by reason and the physical senses, but can create its own inner experiences independent of outer reality.[45]

In his second interlude, Darwin prepares his audience for the dismal despair of the third canto. Although his parries between *Poet* and *Bookseller* are rooted in classical literature, they raise some of the same questions that feature in modern debates about portraying violence. Darwin argues for understatement: subtlety may demand more expertise, but it arouses the deep-felt emotions of tragedy, whereas unmediated horror evokes mere disgust. Far better, Darwin insists, to depict a small bloodstain on a white shirt than to expose the gruesome reality of a thigh shot away by a cannonball. Even Shakespeare and Homer, he contends, were guilty of writing for cheap thrills, whereas Darwin—well, he concludes disingenuously, he'll submit to the reader's judgement.

After the tempestuous third canto, Darwin appears in the guise of a learned philosopher using recent research to debate an old topic—the relationships between the sister arts of painting, music, and poetry. Darwin explores all three sides of the artistic triangle. For the poetry–pictures link, he resumes his earlier theme of under-representation, remarking how moving he found the image of a prisoner's emaciated hand stretched through a grating towards a bowl of soup—far more effective, in Darwin's opinion, than showing his whole body confined to a miserable dungeon. For the other two sides of the triangle, Darwin changes to a technical gear, leaning on contemporary research to compare musical metres with poetic ones, and colours with sounds. In particular, he was intrigued by the potential of an imagined instrument, the ocular harpsichord, which enabled deaf people to enjoy music by producing visible tunes. (Its French inventor might have heard of an older instrument, the cat piano, whose keys drove sharp spikes through cats' tails to generate a melody of yelps; apparently, his employer found the result so hilarious that it completely cured his depression.)[46]

With a last rhetorical flourish as stage-manager, *Poet* urges *Bookseller* to welcome in the audience, who should—he pleads—judge his show tolerantly, as if it were a village play in the local barn. Similarly following his directions, modern historians have treated *The Loves of the Plants* as an amateur production, a frivolous entertainment to be enjoyed for the evening but then forgotten. But if it were truly trivial, the *Anti-Jacobin* satirists would not have paid it so much attention. Having persevered to the end of the poem, I was beginning to understand why they took it so seriously.

When I started reading *The Loves of the Plants*, I had anticipated ploughing through example after example of sexually suggestive encounters between personified flowers. Although I did indeed find plenty of those, only two of the cantos conformed to my expectations: the first, in which the Muse of Botany whimsically describes how plants reproduce; and the last, when Darwin's imagined Goddess offers some well-sugared depictions of the garden as it prepares to close down for the night ahead.

To my surprise, I discovered that the poem varies in both tone and content, ranging over subjects more diverse and substantial than floral flirtations. In several sections, flowers are not so much the focus of Darwin's explanations as the platform for him to hold forth on another topic. Thus in the second canto, Botany retunes her golden lyre to produce 'wilder notes', which celebrate human inventiveness but also hint at a rougher, more tempestuous nature that can destroy as well as nurture. For me, the most extraordinary canto was the third, which opens darkly as deadly nightshade dances around a moonlit graveyard before invading the church to conduct a satanic black mass. Around five hundred grim lines later, Darwin's Muse has been reduced to tears by a catalogue of human iniquities—Medea murdering her children, druidic sacrifices, the warfare that kills mothers as well as soldiers.

By the time I had finished the book, I no longer regarded *The Loves of the Plants* as merely an anodyne frolic through Linnaeus' classification system. Instead, I had formulated an additional interpretation—that its pivotal point is the climax of the third canto. Around forty lines long (411–455), this passage focuses on the evils of slavery, and delivers a potent ideological message. By drawing on plant parallels, Darwin evokes the anguish suffered by mothers whose infants are snatched away to be sold. Recalling how Moses had killed an Egyptian slave-driver and led the Israelites to freedom, he excoriates the British for tolerating the African slave trade. Urging politicians to take responsibility, he points out that those who see but fail to act should also be judged guilty:

> *Hear him*, ye Senates! hear this truth sublime,
> 'HE, WHO ALLOWS OPPRESSION, SHARES THE CRIME.'[47]

I had not expected to reach this conclusion, but from then on, it guided the direction of my research. Instead of following other historians of science by focusing on botany, medicine, and inventions, I would pursue the impression that I had gained from reading the poem for myself, and think about Darwin's attitudes towards politics, slavery, and oppression.

CHAPTER FIVE

Women on Trial

Since, whether sunk in avarice or pride,
A wanton virgin or a starving bride,
Or wandering crowds attend her charming tongue,
Or, deemed an idiot, ever speaks the wrong;
Though nature armed us for the growing ill
With fraudful cunning and a headstrong will;
Yet, with ten thousand follies to her charge,
Unhappy woman's but a slave at large.

Mary Leapor, *An Essay on Woman* (1751)

The next time you hear the expression 'rule of thumb', you might like to reflect on its origins. As late as 1782, a prominent judge voiced only one reservation about wife-beating: the stick should be no thicker than a man's thumb.[48] After marriage, ill-matched couples found themselves trapped in a life that was miserable for both partners, and even in tranquil households, wives were forced into sustaining a subordinate role. Tyrannical husbands and campaigning reformers were alike in comparing the lot of women with that of slaves. One particularly violent and oppressive man declared that 'a wife cannot be kept too much a slave', but even tolerant masters were still rulers.[49] Legal inequalities bred psychological havoc: as Mary Wollstonecraft pointed out, 'slavery will have its constant effect, degrading the master and the abject dependent'.[50]

A marriage proposal was not so much a declaration of love as a business proposition. In affluent families, partners and their parents had to be sure not only that the couple would have enough money to live on, but also that estates were kept intact and dependent relatives cared for. Inequalities were taken for granted. Younger sons were automatically disinherited, and a quarter of them remained bachelors, either because they were trying to avoid the costs of marriage or had been unsuccessful in enticing a rich bride. For a woman, getting married meant being transferred like a piece of property from a father to a husband, handing over her money as well as her legal identity. Remaining a spinster often entailed resignation to a state of dependency on reluctant relatives.

For families able to afford separate lives, husbands' extramarital flings could solidify into semi-permanent relationships, well recognized and perhaps even welcomed by wives who had already borne ten or twenty children. Divorce was extremely rare—averaging fewer than two a year, virtually all filed by men—not because partners were faithful, but because legal action was extremely expensive. As an additional deterrent, any sordid details emerging in court were pruriently picked over in the press by journalists, who used initials to protect themselves against being sued. Guilt was asymmetrical: whereas an unhappy wife needed to prove that her husband's cruelty was life-threatening, a husband had only to prove adultery. In his eyes, the crime was not so much the extramarital sex, but more that a lover had illegally taken advantage of his human property—and there was always the risk that any illegitimate children might lay claim to the family wealth.

These unequal relationships pervade Darwin's *Loves of the Plants*. Even Linnaeus, a pastor's son committed to the sanctity of marriage, had romanticized the fertilization of flowers in a series of unctuous euphemisms. Petals, he wrote, 'serve as bridal beds which the Creator has so gloriously arranged, adorned with such noble bed curtains, and perfumed with so many soft scents that the bridegroom with his bride might there celebrate their nuptials with so much the greater solemnity. When now the bed is so prepared, it is time for the bridegroom to replace his beloved bride and offer her his gifts.'[51] Darwin wrote in a similar vein, but stepped up the sexual innuendo—and stepped outside the bounds of convention.

Sitting in the Rare Books room, browsing through *The Loves of the Plants*, I once noticed that a previous reader had highlighted several verses. I decided to follow this serendipitous clue and see where it led me. According to my anonymous guide, this was the first significant stanza:

Two brother swains, of COLLIN's gentle name,
The same their features, and their forms the same,
With rival love for fair COLLINIA sigh,
Knit the dark brow, and roll the unsteady eye.
With sweet concern the pitying beauty mourns,
And sooths with smiles the jealous pair by turns.[52]

From Darwin's printed notes at the foot of the page, I gathered that he was writing about *Collinsonia*, a herb assigned to Linnaeus' second class of plants because it has two male stamens—the 'brother swains'. During pollination, Darwin explained, the lone female pistil bends over first to one and then to the other male stamen, a sequence he anthropomorphized as a beautiful woman who keeps two lovers happy simultaneously.

How, I wondered, would Darwin fantasize about the mirror situation of men enjoying sex with several partners? I began searching through the book—but without success. Darwin, it appeared, imagined a selective pornographic paradise. All sorts of female stereotypes—the virtuous virgin, the timorous beauty, the laughing belle, the dangerous siren—reflect the desires and prejudices of Georgian gentlemen. Chaste mimosa withdraws from the gentlest touch, seductive grape-vine enmeshes young men in her harlot's clutches, the suffocating breath of West Indian lobelia poisons all around her, and decrepit yet cunning older women pose a threat to unsuspecting youths.

In Darwin's fantasy garden, gods and goddesses might seem to cavort freely, but in reality their amorous exploits are strictly controlled. Although Linnaeus had catalogued many hundreds of species, Darwin personified only eighty-three—and the great majority of those feature just one female with multiple lovers. In fact, he gives only one example of the one-man/several-women scenario, and this is a relatively tame vision of two innocent virgins admiring a handsome young man as he washes his hair, which floats out across the water as a plant.[53]

Female Botanists

Several months later, I spotted the *Collinsonia* lines again—but this time, they were spoken by a woman and reproduced in a comic opera, *The Lakers*, by James Plumptre. When I noticed its publication date—1798, the same year as 'The Loves of the Triangles'—I hoped that I had found a clue worth following.

Plumptre failed to get his *Lakers* staged, although nowadays it is performed from time to time in the Lake District. His bumbling hero is Sir Charles Portinscale, who is trying to evade the clutches of a predatory female botanist, a would-be artistic tourist identified as Miss Beccabunga Veronica of Diandria Hall. In a complicated plot based on deliberately see-through disguises, Sir Charles dispatches his servant Speedwell (Google told me that speedwell is not only a weed, but also a cavern in the Peak District nicknamed The Devil's Arse) to stand in for him and pursue this self-styled Goddess of Botany, thus leaving him free to court her desirable niece. Parodying Darwin, Sir Charles recommends the power of botanical seduction, telling Speedwell 'I have often explained to you the outlines of Linnæus's System of Vegetables; you therefore can very well

> In filmy, gawsy, gossamery lines,
> With lucid language and conceal'd designs,
> In sweet monandrian monogynian strains
> Pant for your mistress in botanic pains.'[54]

(An author's footnote explains that in 'botanic language', 'monogynian' means possessing one female; similarly, 'monandrian' is a pseudo-scholarly term referring to a single man.)

Plumptre was a playwright with a moral mission. A clergyman determined to reform English society, he was appalled at the thought that women might follow the fashion for Linnaeus' sexualized botany. Pompously wrapping his disgust in ornate phrases, Plumptre decreed that botany was 'not altogether a proper amusement for the more polished sex; and the false taste of a licentious age, which is gaining ground, and corrupting the soft and elegant manners of the otherwise loveliest part of creation, requires every discouragement which can be given.'[55] He was far from alone in condemning the new botany. The *Encyclopaedia Britannica* censured Darwin for imagining Collinia with two men. 'A man would not naturally expect to meet with disgusting strokes of obscenity in a system of botany,' it protested, 'But... obscenity is the very basis of the Linnean system.'[56]

Flowers might seem innocently remote from patriotic Christianity, but botany became symbolically relevant during the post-revolutionary unrest of the 1790s. The panicky reactions of upright and uptight Brits remind me of the scaremongering two centuries later, when exaggerated fears about AIDS made it allowable to express in public prejudices against homosexuality that had previously been kept silent. In the wake of the French Revolution—metaphorically the rape of Marie Antoinette—political upheaval and sexual impropriety became interchangeable, so that the health of the British nation depended on cleaning up female morals. As moral hysteria escalated, self-appointed guardians of society insisted that women's behaviour should be reined in still more closely to prevent them from degenerating into the state of promiscuous depravity displayed in Darwin's *Botanic Garden*.[57]

One leading campaigner was Richard Polwhele—like Plumptre, a clergyman as well as a writer, and determined to keep women in their place. At first, Polwhele had admired *The Botanic Garden* so much that he sent Darwin what he called an 'elegant complimental Idyllium' to show his appreciation.[58] But by 1798, the political climate had changed, and Polwhele was a contributor to the *Anti-Jacobin Review*, puffed by the editor for 'employing his poetical talents, in vindication of all that is dear to us as Britons and Christians'.[59] In a savage parody called *The Unsex'd Females*, Polwhele explicitly attacked Darwin. 'Botany has lately become a fashionable amusement with the ladies,' he lamented; 'how the study of the sexual system of plants can accord with female modesty, I am not able to comprehend.'[60]

Although not spectacularly successful at the time, *The Unsex'd Females* has gained new prominence through being absorbed into the canon of feminist literature as an example of male oppression. When I came across the following lines, I immediately recognized their source as the stanza in Darwin highlighted by my unidentified adviser in the library's copy of *The Loves of the Plants*:

> Thrill'd with fine ardors *Collinsonias* glow,
> And, bending, breathe their loose desires below.
> Each gentle air a swelling anther heaves,
> Wafts its full sweets, and shivers thro' the leaves.

> Bath'd in new bliss, the Fair-one greets the bower,
> And ravishes a flame from every flower.[61]

To intensify his parody, Polwhele provided footnotes that dominate the pages. Mostly about women rather than botany, they reveal that the target of Polwhele's lines on *Collinsonia* was not only Darwin but also Mary Wollstonecraft, the enormously controversial author of *A Vindication of the Rights of Woman* (1792). Although Wollstonecraft is now heralded as an iconic founder of feminism, by modern standards her polemical tract is disappointingly timid. She prescribed education for women, not independence. Extolling the virtues of conventional marriage, Wollstonecraft taught that a wife's duty is to care for her husband and her children. Using elaborate botanical imagery, she compared women with force-fed flowers that fade prematurely because their 'strength and usefulness are sacrificed to beauty'. Women, Wollstonecraft insisted, are made, not born—and she blamed men for forcing them to behave more like 'alluring mistresses than affectionate wives and rational mothers'.[62]

Not all women agreed with Wollstonecraft. Hannah More, for instance, bluestocking writer and friend of Samuel Johnson, was a staunch Tory who hated everything French. Her reaction to *A Vindication of the Rights of Woman* might be summarized as 'Rights of Women! We'll be hearing about the Rights of Children next!' In Polwhele's approving descriptions, More features as a Christ-like figure who battles against the devilish Wollstonecraft.

Wollstonecraft's irregular love life gave Polwhele a perfect satirical opening. *Collinsonia*, the flowers with a single female pistil and two male stamens, provided a floral analogue of Wollstonecraft's two disastrous affairs. The first was with the artist Henry Fuseli—not a sensible choice. Far older than her, Fuseli was bisexual but married to a woman who refused to contemplate Wollstonecraft's proposal of a *ménage à trois*. After that romance crashed, Wollstonecraft set off for Paris to recover, but there fell passionately in love with an American adventurer, Gilbert Imlay. This relationship also went very wrong. One illegitimate baby and two suicide attempts later, Wollstonecraft moved in with William Godwin, already notorious as an outspoken and revolutionary political philosopher (and later favourite butt of the *Anti-Jacobin*). At last she was happy, but only for a while: Wollstonecraft died shortly after the birth of her second daughter (who grew up to become Mary Shelley, author of *Frankenstein*).

For a convention-bound provincial clergyman such as Polwhele, this was scandalous behaviour, and—like readers all over the country—he learnt all about it from the *Memoirs* that Godwin published. However well-intentioned Godwin might have been, his intimate revelations proved deeply shocking, and *The Unsex'd Females* was just one of many tirades directed against Wollstonecraft. For example, in 1798, a reviewer for the *Anti-Jacobin* twisted her arguments to accuse her of sanctioning prostitution.[63]

Polwhele portrayed her as Hannah More's satanic opponent, the ringleader of an 'Amazonian band' of 'female Quixotes'—licentious intellectual women touting revolutionary ideals. Accusing them of wearing low-cut dresses like French prostitutes, Polwhele got carried away with his own erotic burlesque:

> With bliss botanic as their bosoms heave,
> Still pluck forbidden fruit, with mother Eve,
> For puberty in sighing florets plant,
> Or point the prostitution of a plant;
> Dissect its organ of unhallow'd lust,
> And fondly gaze the titillating dust;
> With liberty's sublimer views expand,
> And oe'r the wreck of kingdoms* sternly stand;[64]

(*In his long salacious footnote, marked with an asterisk as here, Polwhele claimed that French women were in the habit of biting the amputated limbs of guillotined aristocrats.)

I learnt from Polwhele's title page that he had taken his expression *Unsex'd Females* from Thomas Mathias's *The Pursuits of Literature*, which appeared in instalments during the 1790s. Published anonymously in several parts, this satirical poem had multiple targets but lampooned pretty well anybody who had anything to do with French politics. Within a few pages, an attack on Darwin leapt out at me: Mathias and Plumptre had clearly been reading each other (in the first line, 'cryptogamic' refers to

Linnaean clandestine marriage; the fifth line refers to four men and one woman):

> What?—from the Muse, by *cryptogamic* stealth,
> Must I purloin her native sterling wealth?
> In filmy, gawzy, gossamery lines,
> With *lucid* language, and most dark designs,
> In sweet *tetrandryan, monogynian* strains
> Pant for a *pystill* in botanic pains;[65]

At that stage, I had no idea how revealing other parts of Mathias's tirade would prove to be.

Influence and Plagiarism

Towards the end of *The Loves of the Plants*, in his third interlude, Darwin touched on a delicate topic: how do you distinguish between influence and plagiarism? To borrow from a classical author was standard practice, verging on obligatory reverence, but new laws protecting individual copyright had recently been introduced, and it was still unclear to what extent the phrases of contemporary poets could legitimately be incorporated. To placate Anna Seward, who even pre-publication may have been accusing Darwin of appropriating her verses, he approvingly quoted and referenced a couplet from her elegy to Captain Cook. Well, that's one interpretation. Another possibility is that Darwin boosted Seward's career by writing much of the Captain Cook poem himself, yet allowing it to be published under her name as sole author. Since the surviving evidence is partial (in both senses of the word), it's impossible to be sure.

What does seem sure is that Darwin was worried. He pre-empted criticism by sitting on the fence, acknowledging a few sources and then glossing over the issue with some flowery claims that imported plants would enhance the native beauty of his home-cultivated garden. Whatever the truth, his anxiety does endorse my insistence that Darwin was not some unique eccentric, but was fitting his poetry into a recognized genre. Poetic botany had a long history, and Darwin was not alone in exploiting the erotic implications of flowers.[66]

In addition to whatever concerns he may have had about Seward, Darwin was probably feeling uneasy about some other contemporaries. Reviewers accused him of plagiarizing an earlier Latin poem, *Connubia Florum* (*Marriage of the Flowers*), and it also seems very likely that he had been reading George Crabbe's poetry.[67] Like Darwin, Crabbe was a provincial medical man who studied botany, shot to fame as a poet in the 1780s, wrote in annotated heroic couplets, and included footnoted references about Linnaean classification. On the other hand, he was also a clergyman, he lived into the 1830s, and he wrote in a far less extravagant style. Although older than the Romantic poets, Crabbe was respected by them and remained popular well into the Victorian era.

Some verses of Crabbe and Darwin do seem similar. This is from one of Crabbe's major poems, *The Library* (1781):

> There, like the Turk, the lordly husband lives,
> And joy to all the seraglio gives;
> There, in the secret chambers, veil'd from sight,
> A bashful tribe in hidden flames delight.[68]

And this is from Darwin's *Loves of the Plants*, which appeared eight years later:

> Veil'd, with gay decency and modest pride,
> Slow to the mosque she moves, an eastern bride;
> There her soft vows unceasing love record,
> Queen of the bright seraglio of her Lord.—[69]

You may or may not feel that this provides convincing proof of plagiarism. Whether Darwin was guilty or not, Crabbe certainly resented any implication that their views might be confused. He parodied Darwin in other poems (for instance, 'View that light frame where Cucumis lies spread/ And trace the husbands in their golden bed…'[70]), and he later revised *The Library*, editing out stylistic similarities to *The Botanic Garden* and distancing himself from Darwin's atheistic standpoint.[71]

Whether or not Crabbe influenced Darwin, they both clearly followed the contemporary fashion of looking towards the east for imagery of flowers, romantic love, and enclosed gardens.[72] Four years before *The Loves of the Plants* appeared, William Jones (the Persian scholar) had published his own footnoted poem, *The Enchanted Fruit; or, The Hindu Wife*, a comic verse tale based on the ancient Hindu epic, the *Mahabharata*, which is ten times the length of the *Iliad* and the *Odyssey* combined. One of its key heroines is Drapaudi, the wife of five brothers who are obediently following their mother's injunction to share everything they own equally. Like Darwin, Jones blended botany with mythology, using flowers to naturalize this unusual arrangement (the 'learned northern *Brahmen*' is Linnaeus; the last line refers to many men and one woman):

> Thus Botanists, with eyes acute
> To see prolifick dust minute,
> Taught by their learned northern *Brahmen*
> To class by *pistil* and by *stamen*,
> Produce from nature's rich dominion
> Flow'rs *Polyandrian Monogynian* ...[73]

Influence can operate in both directions, and Darwin also had self-confessed imitators and admirers.[74] Female fans tended to produce sanitized, didactic versions for children, such as these painful lines on the American cowslip, which has five male stamens and one female pistil:

> *Five tender brothers* form'd thy modest train
> And sooth'd my woes with music's heav'nly strain;[75]

In contrast, Richard Payne Knight followed Darwin in taking advantage of botany's erotic possibilities. In this unsubtle example of double entendre, Knight uses floral imagery to comment on the institution of marriage:

> For fix'd by laws, and limited by rules,
> Affection stagnates and love's fervour cools;
> Shrinks like the gather'd flower, which, when possess'd,
> Droops in the hand, or withers on the breast.[76]

On reading this, a half-remembered footnote swam hazily into my mind, and to my surprise I managed to find it. Knight had been accused of plagiarism by Polwhele, and bracketed with Darwin by James Mathias, who disparaged them both as representing 'the decline of simplicity and true taste in this country'.[77] I decided to take another look at Mathias's *Pursuits of Literature*.

In Pursuit of Mathias

I turned straight to the lines where Mathias accused Darwin and Knight of spearheading Britain's decline into bad taste. On a second reading, they still seemed impenetrable:

> In verse half veil'd raise titillating lust,
> Like girls that deck with flow'rs Priapus' bust.
> Go turn to MADAN, and in Gospel truth,
> And *Thelyphthoric* lore, instruct our youth.[78]

Like 'The Loves of the Triangles', Matthias's poem is a complicated cocktail of modern and classical references laced with hatred (or fear?) of anything non-British, non-virile, or non-orthodox. My first breakthrough came when I realized that the second line is about Knight's *A Discourse on the Worship of Priapus* (1786), his study of ancient phallic cults. Although I failed to obtain the book at the British Library, I did manage to find the information I needed.

When you get stuck in a research project, one rescue tactic is to let your mind relax and wander around freely—a sort of free association without worrying about any Freudian implications. I had just been reading about Knight's expedition to Sicily and his friendship with the Dilettante Sir William Hamilton, a British diplomat in Naples. Hamilton was desperately short of money, and thanks to Emma Hamilton's flirtatious interventions (this was before she met Horatio Nelson), Knight helped him out by buying some antique bronzes. From there, I thought back to a marvellous exhibition about Hamilton that I had visited a few years earlier at the British Museum, and remembered my astonishment at seeing a carefully posed arrangement of wax penises. The catalogue answered some of my questions, although it also raised others.[79]

It all started in 1781, when Hamilton sent a letter to his fellow Dilettante Joseph Banks, announcing that while travelling in Sicily, he had 'actually discovered the Cult of Priapus in as full vigour, as in the days of the Greeks & Romans'.[80] Apparently, once a year, women who wanted to become pregnant prayed to St Cosmas, offering him wax models of tumescent male genitalia. And, Hamilton remarked sardonically, all the profits went to the Church. The five examples he presented to the British Museum were tastefully drawn in extraordinary anatomical detail as a still life, and formed the frontispiece for some copies of Knight's book: although some sexual activities were illegal, publishing pictures was not.

For its exhibition, the British Museum used Knight's drawing to create replicas, and I do wonder whether those Italian wax phalluses ever existed. There seem to be so many strange coincidences. Hamilton was apparently never able to witness the ceremony himself because, shortly before his planned visit, the Catholic Church clamped down on such pagan rites. Oddly, there is no proper record of the objects being acquired by the Museum, and the fragile originals are reputed to have crumbled away. Hamilton's jokes make it clear that he was not giving a straightforward account: warning the Museum curator 'to keep his hands off of them', he nicknamed the largest one St Cosmo's big toe. It might all have been a Piltdown-style hoax. But then again, it might not.[81]

Knight's *Priapus* was not simply a disguised sex manual. After all, there is no permanent definition of pornography: Victorian archaeologists covered over the priapic murals in Pompeii, yet Roman parents were presumably relaxed about children seeing them every day at home. Knight's book roused great controversy in Britain, even though in France several respected scholars were, unchallenged, investigating sexual imagery and fertility rituals in ancient religions. His book was misguided rather than wilfully obscene.[82]

In *Priapus*, Knight describes sexual objects originating not only in Greece and Italy, but also Persia, India, Egypt. Most of the illustrations are less salacious (Knight's word, not mine) than his frontispiece, but the prose is deliberately ironic and provocative, packed with complex licentious allusions to near contemporaries such as John Dryden, Edward Gibbon, and Oliver Goldsmith. For instance, Knight compares the Virgin Mary with Diana (Roman goddess of chastity and hunting), equates the dove of the

Holy Spirit with Bacchus (Roman god of wine and drunkenness), and maintains that the Christian cross was originally a phallic symbol of generation. After a concluding diatribe against the Church, Knight's final picture (labelled 'THE END') shows a satyr copulating with a goat.[83]

Knight's mistake was to misjudge his audience. By the 1780s, British readers had become too protective of religious orthodoxy to appreciate the trenchant wit of earlier generations. Originally, Knight's learned research and esoteric humour were intended for restricted circulation amongst gentlemen associated with the Dilettanti Society. It was only when different versions of his book became more widely available that Mathias and his allies started protesting against what they chose to interpret as an attack on British morals and customs. Knight was judged guilty on two scores: he described and illustrated reproductive acts and organs; and he used his discoveries to argue against Christianity.

Despite his book's format, Knight was presenting a serious scholarly study. Paralleling William Jones's theories about language and Darwin's suggestions about evolution, Knight proposed that religions change from one into another over long periods of time. Just as Jones believed that a single ancient language provided a common source for both Asian and European descendants, so Knight maintained that India was the site of an ur-religion based on worshipping creative energy. To support his case, he argued that the Christian Trinity derived from an original creator–destroyer–preserver triad, and that the Greek myths were perpetuating older symbolic roles of the sun and sexual organs. Knight also presented architectural evidence, interpreting Greek flower-topped marble columns as visible reminders of the Indian phallic lingam and self-regenerating lotus. (Nobody then drew the comparison, but these architectural leftovers resemble retained but vestigial organs in biological organisms.) Although he must have regretted the furore he had stirred up, Knight continued to develop these intellectual ideas.[84]

Mathias's next two lines—'Go turn to MADAN, and in Gospel truth, /And *Thelyphthoric* lore, instruct our youth'—proved equally intriguing. I learnt that Martin Madan had started out as a city lawyer, but one evening (perhaps after a few drinks?), he joined a group of friends planning to heckle a sermon by John Wesley. This prank had an unanticipated consequence:

Madan was converted to Methodism and became the chaplain of a hospital for penitent prostitutes. Compared with other European cities, London had an exceptionally high number of street workers, but like many of Madan's contemporaries, I was aghast at his recipe for eliminating the problem—he recommended that polygamy should be legalized.

However shocking or ridiculous Madan's proposal might sound, he certainly took it very seriously. He developed his suggestion at length in the multi-volumed *Thelyphthora, or a Treatise on Female Ruin* (1780–1), which may (or then again, may not) have been a source for William Blake's *Book of Thel*. Madan's *Thelyphthora* (Greek for 'the destruction of women') was a substantial work of scholarship whose erudite footnotes sometimes run over several pages. Justifying his position by citing historical examples and quoting biblical texts, Madan insisted that obliging adulterous men to marry their mistresses would bring two advantages: defenceless women would be financially protected, and sex-starved husbands would be satisfied.[85]

Nowadays, Madan's main claim to fame is that he contributed some lines to the Wesleyan Christmas carol 'Hark the herald angels sing!' His staunchest supporters are Born Again Christians, who—like him—provide ingenious interpretations of biblical texts. Take Jesus's parable about ten virgins waiting for their bridegroom.[86] Five of them wisely planned in advance by making sure their lamps were full of oil, but the other five never quite got round to buying enough supplies: their fate was to remain unmarried. For most interpreters, the moral of this tale is to be prepared for the arrival of Christ; for polygamists, the main point is that one man has acquired five brides simultaneously.

Theologically, Madan was aligned with extreme Protestants and Swedenborgians (followers—including William Blake—of the Swedish Emanuel Swedenborg, a spiritual mystic who experienced a divine call to reform Christianity). According to Madan, since there had been no betrothal ceremony between Adam and Eve in the Garden of Eden, sexual intercourse is in itself a declaration of marriage. Madan held the Catholic Church responsible for many modern evils, including the superstitious cult of virginity and the 'English vice' of sodomy, which he saw as an inevitable consequence of enforced clerical celibacy. In the past, he maintained, both sexes benefited from a polygamous system that ensured effective support for women—but when oppressive Catholicism restricted men to one wife, single women were rendered vulnerable, forced into a life of sin in order to survive.[87]

Ridiculed by his friends, Madan resigned his chaplaincy and retreated into the safer territory of Latin translation. Even so, *Thelyphthora* was neither universally nor immediately panned. For several months, polygamy became a favourite topic in London debating societies, which ran theological discussions on biblical texts such as 'And Lamech took unto him two wives.'[88] Since divorce was prohibitively expensive, many couples separated informally before setting up new relationships that were, strictly speaking, bigamous; Methodists who turned a blind eye to this reality of poorer people's lives were often accused of condoning polygamy. Mary Wollstonecraft defended compassion for fallen women so passionately that she suggested introducing '*left-handed* marriage', in which men would 'be *legally* obliged' to maintain a woman they had seduced.[89] This was the approach Darwin informally adopted. Although he never married his sons' governess, he did bring up their two illegitimate daughters with his other children, and publicly acknowledged their claims on him by buying a nearby mansion so that they could set up their own school.

Unsurprisingly, Madan attracted many critics. One of the most vociferous was his own cousin, William Cowper. His satirical poem *Anti-Thelyphthora* (1781) relates the public shaming of Dame Hypothesis and her ridiculous lover Airy del Castro (Madan), who maintained:

> That wedlock is not rig'rous as suppos'd
> But man, within a wider pale enclos'd,
> May rove at will, where appetite shall lead,
> Free as the lordly bull that ranges o'er the mead...[90]

Mathias's sallies in *The Pursuits of Literature* now made more sense to me, whereas Darwin's *Loves of the Plants* seemed even more complicated but also more interesting. Darwin did not contribute to this public exchange of venom, but he must have felt wounded. Through a chain of family connections, his rejoinder has survived, hidden amongst the papers of the famous statistician Karl Pearson:

Pursuits of letters, what you will,
You'd better lay aside your quill.
You've sure the poetaster's curse,
Who blame bad poetry in worse.[91]

Sexual Politics

Madan's *Thelyphthora* and Wollstonecraft's *Vindication of the Rights of Women* were published respectively at the beginning and end of the 1780s, a decade marked by widespread discussions about marriage, courtship, and suitable behaviour for women. Adultery and sexual violence repeatedly hit the headlines—especially when women seemed to be the guilty partners. In several prominent trials, closely tracked by the newspapers, betrayed husbands sued their wife's lovers for borrowing their property without consent. Although legally in the right, some wronged husbands were awarded only negligible damages by the jury. Perhaps—as with many seminal books—Wollstonecraft was reflecting rather than initiating a more egalitarian approach: she published the right sentiments at the right time, when readers were already prepared to countenance them.

The associations between women, gardens, and sexual temptation went back to the Garden of Eden, and several of these adultery trials prompted salacious double entendres about grottoes and enclosed spaces, about fences and boundary transgression, about the erotic botany of plants. In 1785 a London daily gossip sheet announced that 'Lady Foley and Mrs Arabin have kindly undertaken to plan the intended Shrubbery behind Gower-street—can anyone doubt their *capability*, who reflects with what art they displayed the *beauties of nature* in their own gardens.' Such sniggers referred to the many couples who had been caught illicitly enjoying themselves in shrubberies, a relatively recent innovation in garden design.[92]

The boundaries of an estate symbolized the restrictions on women's freedom imposed by husbands or fathers who ruled over their families as well as over their land. In Jane Austen's *Mansfield Park* (1814), the weak, flirtatious Maria protests that even when the garden is sunny, 'that iron gate, that ha-ha, give me a feeling of restraint and hardship'. On the other hand, virtuous Fanny is careful never to walk alone in the secluded shrubbery, in case she should be trapped and forced into inappropriate behaviour.[93]

Having immersed myself in debates about polygamy and propriety, about power and possession, I had another look at the *Anti-Jacobin*, which—like Mathias—paired Knight and Darwin together. Knight had argued for looser divorce laws by claiming that in earlier societies, marriages could be dissolved at the wish of either party. This parody is from the verse predecessor of 'The Loves of the Triangles', the *Anti-Jacobin*'s 'Progress of Man', which specifically targeted Knight:

> Of WHIST or CRIBBAGE mark th'amusing game—
> The Partners *changing*, but the SPORT the *same*:...
> —Yet must *one* Man, with one unceasing Wife,
> Play the LONG RUBBER of connubial life.[94]

Turning forwards a few pages to 'The Loves of the Triangles', I reread some lines that I'd previously picked out as typically flirtatious:

> Not thus HYPERBOLA:—with subtlest art
> The blue-eyed wanton plays her changeful part;
> Quick as her *conjugated axes* move
> Through every posture of luxurious love,
> Her sportive limbs with easiest grace expand;
> Her charms unveil'd provoke the lover's hand:

This time, the imagery seemed less innocuous, more a barbed attack on feminine wiles than a fantasy about playful seduction. I continued to the next verse, which sardonically invites Ellipsis to secure her future happiness by luring her man into the permanent trap of marriage. Suddenly, the mood darkens and the scene switches to revolutionary France. It was becoming ever clearer to me that 'The Loves of the Triangles' was not the light-hearted sexual romp I had been led to believe.

CHAPTER SIX

Seraglios

As soon as I entered, the music and dancing stopped. The assembled ladies and gentlemen thronged round me in wide-eyed amazement and examined my robe, turban, shawl and other parts of my costume.... How ironic that I, who had gone there to enjoy a spectacle, became a spectacle myself.

Mirza Sheikh I'tesamuddin, *The Wonders of Vilayet: Being the Memoir,
originally in Persian, of a Visit to France and Britain in 1765*

When James Cook sailed back from his second trip to the Pacific, he brought with him Omai, a Tahitian hoping to rally British support for his own domestic political campaign. Confounding his expectations, far from being treated as an aspiring ruler, Omai became a high-society mascot, a favourite dinner guest condescendingly appreciated for his quaint accent and elaborate tattoos, but mocked for being flummoxed by ice and magnifying glasses. I was pleased to learn that he regularly defeated his aristocratic acquaintances at chess and backgammon. On his departure, they showered him with gifts bound to come in handy on a Pacific island—a custom-made suit of armour, an electrical machine, and some furniture.

Viewed in retrospect, the arrogance and self-confidence of eighteenth-century Europeans seems extraordinary.[95] And that applies to women as well as to men: Mary Wollstonecraft may be a feminist figurehead, but she was no great supporter of sisterly solidarity. Like her male contemporaries, she regarded other cultures as inherently inferior. In her opinion, not even the lowest of European women should allow themselves to behave like the subjugated, mindless creatures of the east. British spinsters desperate to get married, she wrote, 'act as such children may be expected to act—they dress, they paint, they nickname God's creatures. Surely these weak beings are fit only for a seraglio!'[96] (By this time, 'seraglio' and 'harem'—both appearing in a variety of spellings—referred either to a group of Muslim wives or to the rooms in which they lived.)

As travel and trade increased during the second half of the eighteenth century, so too did British interest in other cultures. Britain was turning eastwards, away from her lost American colonies, in order to concentrate on developing links with Asia and Africa. Commercial exploitation and government expansion became almost indistinguishable, since private companies were heavily involved in political administration. India, for instance, was largely controlled by employees of the East India Company; stationed many weeks away from London by ship, they were effectively in charge of local affairs.

Globalism first became significant as an economic concept in the 1990s, but Britain had already been linked into international networks over two centuries earlier. Omai probably sipped Indian tea sweetened with Caribbean sugar out of Chinese porcelain dishes. Like Omai himself, many imports seemed simultaneously fascinating and repellent, noble and savage, exotic and bathetic. In *The Loves of the Plants*, Darwin catered for his readers' desire to gaze at the unknown while staying safely behind British doors. As Horace Walpole put it, Darwin had 'been inspired with such enthusiasm of poetry by...peeping through the keyholes of all the seraglios of all the flowers in the universe!'[97]

Accompanying greater familiarity with other civilizations came widespread if unspoken concerns about British superiority. Was it perhaps possible that another society could be different without being inferior? One way of refuting that subversive suggestion was to insist that the British way of life simply must be the best: after all, look what had happened during the Reign of Terror across the Channel. A less hysterical approach to ensuring that British remained best was to concoct a theory of human variation based on nature—a scientific law vindicating inbuilt assumptions by explaining why north-west Europe was home to the most advanced cultures in the world. Equatorial climates, this argument ran, not only make people dark-skinned and lazy, but also increase their sexual desire. In contrast, those living towards the poles are likely to be pale, hard-working, and frigid. Temperate zones such as Europe owe much to the relative restraint of both the weather and the women, who are perfectly fitted to their submissive roles as wives and fertile mothers.

This type of environmental philosophy gratifyingly reinforced British superiority: 'The burning ardours, and the torturing jealousies, of the seraglio and the haram, which have reigned so long in Asia and Africa' stand in stark contrast to the northern 'spirit of gallantry, which employs the wit and the fancy more than the heart'.[98] In Figure 3, Gillray's caricature conveys this judgement visually, showing British politicians enticed by voluptuous and voluminous female slaves. In a picture on the wall, half-naked women lounge indolently beneath a palm tree, while the book on the table is a pornographic biography of the erotic poet and rake, John Wilmot, Earl of Rochester.[99]

Historians suggested that humankind had developed from early hunting societies into settled farming communities and then into commercial civilizations. Because these transformations occurred at different rates,

Figure 3 James Gillray, *Philanthropic Consolations, after the Loss of the Slave Bill.* Hand-coloured etching with engraving (4 April 1796). © Trustees of the British Museum.

places and peoples encountered on voyages overseas were regarded as far away not only in distance but also in time: to be non-European was also to be backward and primitive. One way of measuring the level of a society was by the status of its womenfolk, and in the eyes of European gentlemen, their own had advanced further up along the human scale than anywhere else in the world—their wives, sisters, and mothers led more leisurely lives, were better educated and received greater love from their men. From there, it was an easy rhetorical step to pronounce that since women already enjoyed far more privileges than their predecessors as well as their counterparts overseas, they should stop campaigning for more, and instead be content with their lot.

Some Europeans were more equal than others. Englishmen confidently placed themselves at the top of the hierarchy, looking down on inhabitants not only of the Mediterranean countries, but also of France and of Scotland. Amongst women, aristocrats came top, followed by the middle classes, and then servants and prostitutes, who themselves lay above Asians and Africans. Such prejudices persisted well into the nineteenth century. When Charles Kingsley, author of *The Water Babies*, travelled round Ireland, he described how disturbed he was by the 'human chimpanzees' he met—'To see white chimpanzees is dreadful,' he complained; 'if they were black, one would not feel it so much.'[100] Irony perhaps, but presumably he thought it was witty.

In a published talk first addressed to the Literary and Philosophical Society of Manchester, a surgeon called Charles White posed a comforting rhetorical question: 'Where', he asked, 'shall we find, unless in the European, that nobly arched head, containing such a quantity of brain...?' When it came to women, however, White shared with Darwin a greater interest in their sexual than their intellectual qualities. 'In what other quarter of the globe shall we find the blush that overspreads the soft features of the beautiful women of Europe, that emblem of modesty, of delicate feelings, and of sense?.... Where, except on the bosom of the European woman, two such plump and snowy white hemispheres, tipt with vermillion?'[101]

The position of European women relative to dark-skinned men was not clear-cut, and comparisons between non-Europeans, animals, women, and slaves were common. Jokes do not, of course, necessarily represent an individual's true feelings; nevertheless, they only work because they reflect real attitudes. For Samuel Johnson, a woman preaching resembled a dog walking on its hind legs—although not done well, you are surprised to find it done at all. Similarly, when David Hume heard of a learned black Jamaican, he commented that 'it is likely he is admired for slender accomplishments, like a parrot who speaks a few words plainly'.[102] Slaves were grouped into categories. Those bought for hard labour—mainly male Africans and Tartars—were denigrated as being ugly and stupid, whereas women from the Caucasian regions (often acquired for the luxury end of the sex market) were idolized as epitomes of human beauty. When nineteenth-century anthropologists divided human beings into separate races, they inherited this label of 'Caucasian' with its inbuilt implications of whiteness, beauty, and superiority, and extended it to cover a wide geographical range.[103]

Looking back at Darwin's erotic botany, trying to place myself in the position of a contemporary reader, I wondered whether it might be less intrinsically demeaning than I had originally thought. As Walpole noticed, at least he appeared to have given women a dominating role: 'the universal polygamy going on in the vegetable world...is more *galant* than amongst [the] human race, for you will find that they are the botanic *ladies* who keep harems, and not the *gentlemen*'.[104]

On Being Foreign

When I first encountered Martin Madan and his *Thelyphthora*, I had two reactions. I felt excited at discovering a new way of thinking about *The Loves of the Plants*, but I also wondered at Madan's naivety in believing that making polygamy legal would solve the problems of prostitution. The more I read, the more I recognized my own naivety. Like an Enlightenment explorer gradually realizing that the Africa imagined by Europeans was very different from the Africa of actuality, I began to appreciate that as I wandered about the foreign land of the eighteenth century, I was bringing the wrong preconceptions to polygamy.

The very word 'polygamy' (Greek for 'many women') indicates that this was an asymmetrical debate: the reverse—polyandry—was not an issue.

A woman with more than one husband was, spluttered Madan, 'too abhorrent from *nature*, *reason*, and *scripture*, to admit of a single argument in its favour, or even to deserve a moment's consideration'.[105] Even those willing to consider polyandry promptly rejected it on economic grounds, pointing out that the identity of a child's father (in other words, its financial supporter) would be uncertain.

Rather than being an isolated eccentric, Madan was engaging in an ongoing discussion whose protagonists included such eminent figures as the philosopher David Hume and the novelist Samuel Richardson.[106] I was surprised to discover that Madan was not the only clergyman amongst the defenders of polygamy, for whom it seemed less scandalous than divorce—the reverse of what many British people feel today.

Often, the question was not so much whether polygamy should be allowed, but more about how an existing practice could be justified. Because divorce was so expensive, many separated couples never split legally, so that polygamy referred not only to enjoying several women simultaneously, but also to living permanently with a second partner, a practice now regarded as serial monogamy. According to Hume, the Turkish ambassador to France commented that Christians were more sensible than Muslims, because they economized by enjoying seraglios in their friends' houses instead of paying for their own. Although he was being sarcastic, there would have been no point in the comment if there were no truth behind it.

One historical tourist's blunder I made was paying insufficient attention to the Bible. The satirist Daniel Defoe remarked that if God had intended men to be polygamous, He would have created companions for Adam from six of his ribs, not just one.[107] Understanding this joke entails realizing that polygamy was invoked in religious controversies. Extreme Protestants endorsed the custom by citing its frequent occurrence in the Old Testament: as just one example, the book of Chronicles reports that King David had eight wives (although perhaps a more accurate modern translation would be 'eight legally recognized sexual partners'). In contrast, others denied that the scriptures should be interpreted literally and unquestioningly. Pointing to the New Testament authors who insist on monogamy, they argued that either God had contradicted Himself—a clear impossibility—or else the Bible is internally inconsistent.

On top of these biblical niceties, there were two practical arguments in favour of polygamy: sex and money. The sexual deprivation line of thought blatantly protected male interests, and was (supposedly) justified by the growing middle-class fashion for breastfeeding. Instead of employing wet nurses, wives were urged to nurture their babies themselves—a recommendation that conveniently encouraged them to focus on their homes and families rather than on careers and campaigns. Unfortunately, this enhanced maternalism precluded sexual activity, which was said to spoil a mother's milk. Although nobody seems to have worried about the women, husbands claimed that they needed to alleviate their feelings of frustration during the frequent post-pregnancy periods.

Economic arguments were crucial in both divorce and polygamy. Hume, for example, took it for granted that marriage is a contract entered into for generating children, so that husbands are responsible for supporting their families.[108] In contrast, when single women were unlucky enough to become pregnant, they were regarded as social outcasts and risked starving to death. Although Madan and Wollstonecraft might well have objected to my bracketing them together, they did both face up to reality in maintaining that unmarried mothers should be financially supported by their lovers.

In Madan's eyes, Britain was the new Orient, whose public brothels were shameful seraglios.[109] Yet despite his compassion for prostitutes, not all women earned his charity. Like so many British people, although Madan must have read about the appalling treatment meted out to black slaves in the Americas, he apparently felt it a wrong not worth acting upon—the chapel he built at the Lock Hospital to save London's fallen women was funded by the profits from his Caribbean plantations. Evangelicals were more concerned to save souls than to save slaves, and wealthy philanthropists judged that being enslaved in a Christian institution was preferable to freedom without faith.[110]

Conveniently for British men overseas, travellers' accounts condoned treating dark-skinned women differently from the paler family members left behind at home. African and Asian women were seen as doubly inferior, marred not only by their dark skins but also by the torrid climates that exacerbated their already overheated constitutions. Settlers reported that in many places polygamy was fully sanctioned as an organized system that made good economic sense. Although indigenous husbands often had favourites among their partners, they cared for all their women and offspring, while essential tasks such as growing food, looking after children, and satisfying men's sexual needs were shared by a large interrelated labour force. Until zealous missionaries started preaching that monogamy is an essential characteristic of civilization, visitors were prepared to ogle and envy, not automatically condemn. Indeed, many of them went further, taking multiple wives for themselves, or chauvinistically assuming that local seraglios operated as brothels.

Unsurprisingly, back in Britain, there were mixed attitudes towards mixed marriages. Asians as well as Africans were regularly referred to as blacks, but Africans were regarded as more threatening. Although British patriots declared that the purity of their island people would be contaminated by black Africans, they were less concerned about Anglo-Indians fathered by white settlers overseas. Whereas it was acceptable for a white man to return from overseas with a black bride, a black husband with a British wife was condemned as a predator—and the white woman was despised as the most sinful since Eve. When Darwin discussed mixed marriages in *Zoonomia*, he simply assumed without question that the wife would be black and the husband white—and drawing an analogy with mules, he inferred (with questionable logic) that the father would have a stronger influence on the children before birth.[111]

When I read travellers' accounts of their experiences in India and Asia, I sometimes feel as if I were entering an imaginary world created by male fantasy. Did a shipwrecked sailor really survive because so many women begged him for sex that he generated enough children to cultivate his desert island? Why did indigenous mothers not experience any pain when they gave birth? Could it be true that dark-skinned females were aggressive seducers who accosted white strangers in public, stripping off their trousers and demanding to be gratified? Did African parents regularly kill, sell or even eat their own children? Such tales of power, prejudice, and possession remind me of those allegorical frontispieces in which the continents are symbolized as dusky alluring women inviting European explorers to map and conquer them.

Conversely, while these supposedly factual accounts reveal much about their authors' imaginations, those marketed as fiction can be informative about what was actually going on. In *The Indian Adventurer* (1780), Mr Vanneck claims moral superiority by rescuing oppressed women from the extended harem of a tyrannical nabob, yet has no compunction in reaping the maximum number of financial rewards. What seemed exploitative opportunism to me made commercial common sense for Vanneck: he bought slaves as presents to win favour with European women, trawled the streets with a friend to take home cheap fun for the night, and gloated over the valuable jewellery given to him by a wealthy Indian woman. His unabashed note of triumph suggests that this was standard behaviour.[112]

Otherness works both ways. Although Englishmen felt that their advanced state of civilization justified them in enjoying the delights of hot climates, their victims naturally resented the disruptive influence of these foreign invaders. Explorers brought with them not only Christianity but also sexually transmitted diseases and guns. Subordinated at the time, indigenous people have been further suppressed by being silenced in the historical record. Often illiterate, perceived by Europeans as a homogeneous crowd rather than unique individuals, they have left little concrete evidence either in their own voices or through those of their victors.

One rare exception is Ignatius Sancho, who claimed to have been born on a slave ship. Eventually taken in by an enlightened British family, he subsequently achieved independence as a London shopkeeper and became England's first black voter. Benefiting from the education he had received, Sancho castigated traders for encouraging petty despotism in Africa. 'The grand object of English navigators—is money—money—money,' he insisted; 'the poor wretched natives—blessed with the most fertile and luxuriant soil—are rendered so much the more miserable for what Providence meant as a blessing:—The Christians' abominable traffic for slaves— and the horrid cruelty and treachery of the petty Kings—encouraged by their Christian customers...'[113]

From Sancho's perspective, it was Christianity, not Islam, that was the oppressive religion. Similarly, although Wollstonecraft accused

Muslims of holding women captive, a Turkish wife believed that 'The Husbands in England were much worse than in the East; for that they ty'd up their Wives in little Boxes, of the Shape of their bodies.'[114] To a fashionable English lady, corsets seemed indispensable for making her attractive; to the member of a harem, being laced into tight underwear appeared an intolerable restriction on her freedom. Even the fabricated label 'seraglio' resonates with prejudiced assumptions, because it conflates the Turkish word for palace (*serāī*) with an Italian word (*serraglio*) that sounds similar but means something very different—a place of confinement.

Exoticism

When I came across Sophia Watson, I was intrigued. Concealing her true identity, in 1768 she published *Memoirs of the Seraglio of the Bashaw of Merryland*, a title suggesting that hers was far from being a sober travel narrative. Although I was unable to find out much about Watson, I do know that she had a vivid imagination. In her book, she features as an aggrieved sultana who has decided to expose a pasha's sexual games. After supposedly studying at first hand how such things were organized in Turkey, the grandee had established his own seraglio in the imaginary country of Merryland, examining prospective recruits by gazing at them secretly through a small latticed opening. I immediately thought back to Walpole's comment about Darwin peeping through the keyholes of flowery seraglios. Later, I learnt that this voyeuristic ritual originated not in Turkey, but inside the romances of authors titillating British readers whose personal harems existed only in their own minds.[115]

The Bashaw's inspection grille illustrates how Eastern reality was converted into Western fantasy. Watson's Bashaw had no first-hand experience of Turkish culture, but took what he fancied from books by travellers who themselves had only a superficial knowledge. The Enlightenment fashion for all things Eastern started at the beginning of the eighteenth century, when *The Arabian Nights* was first published in Europe—or to be more accurate, Westernized versions appeared. Translated from Arabic into Latin into French into English, Britain's *Arabian Nights* included several stories—Aladdin and his lamp, Ali Baba and the forty thieves, Sinbad the sailor—that were not in the original collection.

Another influential if not entirely reliable source of information is the correspondence of Lady Mary Wortley Montagu, an adventurous woman who travelled around Turkey for some months as the British ambassador's wife. Perhaps her most valuable Turkish import was smallpox inoculation, which she introduced in the 1720s on her return to London. Forty years later, after a complicated period mostly spent in self-enforced exile pursuing various lovers, Montagu published her embroidered descriptions of harem life, which sparked off a fresh craze for exotic costumes and furnishings.

Claiming that as a woman she had access to the innermost reaches of seraglios, Montagu put herself in the Bashaw of Merryland's position by providing eyes for male artists: 'I had wickedness to wish secretly that Mr Gervase [the portrait artist Charles Jervas, who had painted her as a shepherdess] could have been there invisible...so many fine Women naked in different postures, some in conversation, some working, others drinking Coffee or sherbet, and many negligently lying on their Cushions while their slaves (generally pretty Girls of 17 or 18) were employ'd in braiding their hair...'[116] Even though she appeared to be a reliable correspondent, Montagu was elaborating on actuality: according to Victorian women who penetrated behind the screens, harem life was nowhere near as indolent and sexually charged as she made out.

Montagu's versions of Turkish poetry were also far removed from the originals. Like many other early translators, she possessed only limited linguistic skills, and focused instead on rewriting traditional legends in the heroic couplets so fashionable in England. That attitude towards Eastern literature changed dramatically in the last quarter of the eighteenth century, when scholars started to become fully fluent in languages such as Persian, Arabic, and Turkish. Academic fashions rarely emerge in isolation. Rather like the explosion of gender studies sparked off by the women's liberation movements of the 1960s, scholarly studies of the East were stimulated during the Enlightenment by Britain's commercial and imperial expansion. In particular, Persian benefited from being the official language for Indian diplomats. Schools were set up to train civil servants, information about Asian culture was published in English-language journals, and printing presses with Persian characters were built in both India and Britain.

Amongst this new generation, the most important individual was William Jones, who possessed sufficient insight and courage to remark—long before the protests of Edward Said—that the 'Orient' is 'a word merely relative'.[117] Factually, what lies to the east depends on where you're looking from—as an Australian prime minister remarked, 'What Great Britain calls the Far East is to us the near north.'[118] Conceptually, from a chauvinistic British perspective, anything oriental was romanticized into seeming simultaneously exotic and inferior, alluring and threatening, civilized yet dangerously erotic. Imported tea, china, and silk were physically real, but the societies described in oriental tales were products of Western imaginations.

Jones wanted not only to give Europeans the opportunity of enjoying Eastern literature, but also to demonstrate that it rivalled or even surpassed the great European classics—Homer, Ovid, Virgil. His admirers insisted that, despite the ornate imagery and fanciful myths, Persian gins are not so very different from British fairies or Greek goddesses. To make this potentially alien culture acceptable, Jones avoided literal translations, instead converting both prose and verse into a familiar poetic form such as rhyming couplets, even though that meant sacrificing the metrical structure and rhyme schemes of the originals. Another stratagem he adopted was to juggle with gender. In Persian, the same word can mean either 'he' or 'she', and for his British readers Jones disguised homoeroticism by making it appear that the recipients of love verses were women.[119]

Darwin and Jones were pre-Romantic transitional figures, products of their upbringing who initiated their own demise by transforming rather than replacing what they had inherited. Despite this similarity, they portrayed Eastern cultures very differently. Jones was determined to overturn wishful-thinking views of Islamic life. In 'The Enchanted Fruit', his poetic version of scenes from the *Mahabharata*, he reversed the direction of prejudice by presenting Indian women as the paradigm of virtue: for Jones, white did not necessarily mean good. These words are spoken by the goddess Britannia as she defends British women against charges of immorality:

> What! are the fair, whose heav'nly smiles
> Rain glory through my cherish'd isles,
> Are they less virtuous or less true
> Than *Indian* dames of sooty hue?[120]

In contrast with Jones's informed depictions, Darwin fell back on stereotypes. For instance, he orientalized one plant as a 'Turkish lady in an undress', thus preserving British virtue by tying together eroticism and exoticism.[121] Describing in a long footnote how Mimosa readily closes its leaves for protection, he (perhaps following Crabbe) imagined it—or rather, her—as a chaste, timid Muslim:

> Veil'd, with gay decency and modest pride,
> Slow to the mosque she moves, an eastern bride;
> There her soft vows unceasing love record,
> Queen of the bright seraglio of her Lord.—[122]

Unlike Jones, a real traveller, Darwin was untroubled by finer points of geography. Unless you pay attention, it's not immediately obvious which part of the world is being exploited for its exoticism. Only a few lines later, he transports his readers to the African desert, where dying panthers and writhing serpents surround 'Fair CHUNDA' with her 'brow unturban'd... rising bosom and averted cheek'.[123] Soon after encountering a swan with a black crest and a spotted pard (a legendary and lethal big cat that mates with a lion to produce a leopard), readers suddenly arrive in Tahiti, where the grand finale of a wedding between a hundred couples is taking place.

Accounts of Cook's voyages and satires on Banks's escapades had characterized Tahiti as a paradise of free love, an oceanic Arcadia uncorrupted by constraints of civilization such as compulsory monogamy. British astronomers recruited local inhabitants to help them build Fort Venus, supposedly named after the planet they were observing: their diaries make it clear that they were more interested in the alternative interpretation. Four years before *The Loves of the Plants* was published, the Theatre Royal in London put on a sell-out Christmas pantomime inspired by Omai's visit to Britain. Although in reality Omai's return to Tahiti had been ignominious, onstage he starred as a splendidly costumed, autocratic ruler. Towards the end of the play, a mad prophet declared that King Omai was 'the owner of fifty red feathers, four hundred fat hogs, and the commander of a thousand fighting men and twenty strong-handed women to thump him to sleep'.[124] Darwin sprinkled his own description of Tahiti with similarly loaded terms—promiscuous, faithless, licentious, meretricious. This can only have reinforced his readers' conviction that the Pacific Islands were, like Asia and Africa, havens of opportunity for sex-starved Europeans.

One reviewer commented that Darwin wrote with 'an oriental luxuriance of imagination'.[125] When I looked up the relevant lines, I saw what he meant. Darwin clumsily imitated the Turkish floral language of love to summon up visions of enclosed gardens, luscious flowers, and unattainable exotic women:

> So, when the Nightingale in eastern bowers
> On quivering pinion woos the Queen of flowers;
> Inhales her fragrance, as he hangs in air,
> And melts with melody the blushing fair...[126]

Darwin's oriental nightingale came from a famous Persian legend first introduced into England by Mary Wortley Montagu, and later—under Jones's influence—translated properly. With its theme of unrequited love, the tale of the rose and the nightingale became a Romantic favourite, repeatedly translated and retold, perhaps most memorably by Oscar Wilde. In his heart-rending if misogynistic version, a nightingale overhears a student longing for a red rose to woo his professor's daughter. To gratify his wish, the bird impales herself on a thorn and sings all night while her blood pulses out, staining a white rose red. Delighted, the student presents the flower to the girl, but she brutally rejects him in favour of a richer suitor.

Unusually, Darwin gave this moral fable a happy ending by omitting the bird's fruitless self-sacrifice:

> Admiring Evening stays her beamy star,
> And still Night listens from his ebon car;
> While on white wings descending Houries throng,
> And drink the floods of odour and of song.[127]

Darwin's facile orientalization came under fire from the *Anti-Jacobin* satirists. In the second instalment of their 'Loves of the Triangles', they introduced poetic stage props such as Aladdin's lamp, a satanic magician and 'GINS—black and huge!' The accompanying mock-scholarly footnotes cited a *New Arabian Nights* by Cazotte, the author of *Le Diable amoureux* (*The Devil in Love*). My immediate response was to take this as a joke, but after conscientiously looking him up, I discovered that Jacques Cazotte really did exist—before, that is, he was guillotined in the Revolution. To make things still more complicated, Cazotte's works are themselves pastiches with political messages. How many layers of irony and self-reference must lie within 'The Loves of the Triangles'?[128]

Reading on, I felt myself becoming enmeshed in the parodists' 'enchanter's hand' that mathematically traces out a 'mystic *Circle*' with a '*Hypothenusal* wand'. At some point, I felt, I just had to decide that enough is enough, as otherwise this research project would simply spiral on and on, or—worse—down and down. Flicking through 'The Loves of the Triangles' for the umpteenth time, I recalled that the very first stanza refers to Darwin's second major poem, *The Economy of Vegetation*. Resolving to abandon *The Loves of the Plants* for the time being, I turned to what would become (confusingly) the first part of *The Botanic Garden*.

THE ECONOMY OF
VEGETATION

KNOWLEDGE, POWER, AND
SOCIETY

Once in a lifetime
The longed-for tidal wave
Of justice can rise up,
And hope and history rhyme.

Seamus Heaney, *The Cure at Troy* (1990)

I drifted awake to hear the radio informing me that the best place to look for extraterrestrial life is on Goldilocks stars—neither too hot, nor too cold, but just right. Since that's a bedtime story I've told many, many times, I decided to look up its origins, and was surprised to discover that it first appeared in print as late as 1837. The original published version was by Wordsworth's close friend and fellow Lake poet, Robert Southey. Together with Coleridge, as young men they had dreamed of setting up a utopian community. Although they never quite got round to organizing it, Southey described the next best thing—three bachelors who lived happily together in the woods before being disturbed by an ugly old woman, who jumped away through the window, never to intrude again into this harmonious collective. (This witch-like character was replaced twelve years later by a more comforting golden-haired girl.)

Wordsworth had already written his own fairy tale about the countryside, the poem 'Goody Blake and Harry Gill'. He credited his source—Erasmus Darwin. In *Zoonomia*, Darwin described an ancient witch who was illegally gathering firewood: when the farmer caught her red-handed, she put a curse on him, declaring that he would never again be warm. For the next twenty years, the farmer lay wrapped up in bed, psychologically convinced that he would die of cold should he venture out. Wordsworth recast Darwin's medical anecdote as a poetic fable, delivering a political message that drew attention to the plight of the rural poor.

The same year that Wordsworth published this parable in *Lyrical Ballads* (1798), similar imaginary characters—the traditional Cinderella, and also Aladdin, recently imported in *The Arabian Nights*—featured in a very different political poem: 'The Loves of the Triangles'. By the end of this satire, what starts out as a fairy kingdom has been transformed into an establishment nightmare—a demonic republic whose 'Imps of Murder' are building a ship, a '*Floating Frame*', for Napoleon's troops to cross the Channel and invade Britain. This imagined military manoeuvre was inspired by some recent real naval mutinies, when British sailors' demonstrations in the docks had been blamed on Jacobin infiltrators. Embroidering their scenario, the *Anti-Jacobin* described miniature 'Sylphs of DEATH' who clap their 'tiny hands', tittering and laughing as they fire off mortar shells before strapping Prime Minister Pitt to the plank of a guillotine and watching his 'liberated head' roll away.

While the anti-Jacobins were weaving these and other panic-inducing fantasies, their radical opponents were generating their own exaggerated tales—the anti-anti-Jacobin propaganda, if you like. Years before he conjured up Goldilocks, Southey compiled *Letters from England* (1807), a book exploiting the literary trick of disguising himself as a foreign tourist bewildered by bizarre British customs. Browsing through Southey's satire, my eyes fine-tuned for references to anti-Jacobinism, I was struck by this passage:

> During one period of the French Revolution the Brutus head-dress was the mode, though Brutus was at the same time considered as the Judas Iscariot of political religion, being indeed at this day to an orthodox anti-Jacobine what Omar is to the Persians; that is, something a great deal worse than the Devil. 'I suppose, sir,' said a London hair-dresser to a gentleman from the country,—'I suppose, sir, you would like to be dressed in the Brutus style.' 'What style is that?' was the question in reply. 'All over frizzley, sir, like the Negers,—they be Brutes you know.'[1]

Fascinating. But what did it mean? What had Brutus and Omar (and anyway, which ones?) got to do with the French Revolution? However obscure this seemed to me, I knew that Southey made his readers laugh—which is why his book sold so well. Whereas they understood his references immediately, it was only after a good deal of reading and Googling that I felt I was even beginning to get the joke.

I imagine that many Muslims reading this will already have identified Omar as the seventh-century Caliph who enlarged the Islamic empire, conquered the Persians and ordered them to convert. Judas I recognized as the Apostle who betrayed Jesus by selling him to the Romans, and Marcus Junius Brutus as the Roman politician who betrayed Julius Caesar by killing him. This Brutus modelled himself on an earlier one—Lucius Junius Brutus, who established the first Roman republic in 508 BCE by overthrowing the tyrannical King Tarquin. When Lucius Junius discovered that his own sons had gone over to the royalist side, he not only ordered them to be executed, but also watched them die.

During the French Revolution, this paternal sacrifice became a symbol of devotion to the state. In 1789 the artist Jacques-Louis David painted a massive, stylized canvas, *The Lictors Bring to Brutus the Bodies of His Sons*, in which Brutus turns aside as his wife and daughters reach out their arms towards bloody corpses being carried by on stretchers. The following year, Voltaire's play *Brutus* was revived in Paris. On the first night, the audience split into two camps—royalists who cheered the autocratic King Tarquin, and radicals who broke into applause when Brutus proclaimed: 'Gods! Give us death rather than slavery!' On the second night, in a planned political demonstration, only Republicans were present as the cast recreated David's tableau on the stage. The actor playing Brutus emphasized his break with the past by abandoning the conventional stage costume of frock coat and ornate powdered wig, instead appearing as a Roman in loose robes and sandals, his dark cropped hair brushed forwards. Overnight, Brutus became an icon of Revolution, his image on sale as busts or engravings, his name a favourite for newborn babies. To show their support, Republicans—women as well as men—adopted natural, flowing clothes and hairstyles.[2]

(At that point, I accidentally opened a book at the wrong page and discovered that hair was a sensitive political topic in 1790s Britain, because Pitt had levied a tax on hair powder to raise funds for his military campaigns.[3] I had, and still have, no idea whether there was any connection, but it seemed a clue worth noting. Projects often trail loose ends behind them, and I welcome suggestions.)

'Brutes' was clearly a wordplay on 'Brutus'—but what about the reference to 'frizzley' hair? Some time in the middle of the night (research never leaves obsessive academics alone), I remembered a curious caricature of a black-and-white man split down the centre (Fig. 4). Published in 1792, it was drawn by Richard Newton, only fifteen years old but already famous for his anti-slavery images. Here was visible evidence that British people associated French revolutionaries attacking aristocratic oppressors with Africans rising up against their owners.

Figure 4 Richard Newton, *A Real San-Culotte*. Hand-coloured etching (1792). © Trustees of the British Museum.

Newton's immediate inspiration was an insurrection in the French colony of Saint Domingue, which had begun the previous year. Led by 'the black Napoleon', François-Dominique Toussaint Louverture, it lasted until 1804, when Haiti became the first and only republic to be founded by slaves who had rebelled. Saint Domingue's plantations produced 40 per cent of the world's sugar and brought huge profits to the Europeans who ran them, although the enslaved Africans endured atrocious conditions. After the island's planters proposed breaking away from France, the workers were—with justification—terrified that without central restraint, colonial oppression would intensify still further. On both sides, the violence was appalling. For wealthy Brits at home, the implications were clear: the spirit of revolution might creep not only over the Channel, but also across the Caribbean to infect slaves on plantations run by their own friends and relatives, who would be massacred and ruined like their French neighbours.

On the right of Newton's caricature is the devil, painted the same deep bluish grey as he used for African and Caribbean slaves. On the left is a French revolutionary dressed in red, blue, and white, the smile on his face belied by the bloody dagger in his hand and the tools in his belt. He is immediately identifiable as a radical by his *bonnet rouge* (red cap), the French version of a Phrygian hat. (Originating in Turkey, this symbolized liberty for freed slaves in the Roman empire, and had more recently been adopted by American colonialists seceding from the British.) The rebel's bare leg indicates that he is a *sans-culotte* (no breeches), a style adopted by Parisian labourers to show their contempt for the philosophical revolutionaries who wore breeches made from velvet presented to them by wealthy salon hostesses. In Britain, *sans-culotte* resembled *Jacobin*, a term of abuse used indiscriminately to signify a violent revolutionary.[4]

Could it be that beneath the skin there was little to distinguish an African slave from a lower-class European? This motif of a divided figure features in several images of the time, usually to imply that binary oppositions—man/woman, Tory/Whig, rich/poor—might not be so clear-cut after all. Although the French and American Revolutions are now mostly remembered in separate national histories, towards the end of the eighteenth century they were often referred to as a single movement. The black devil's cloven hoof is planted not in the Caribbean but in North

America, whose colonies had only very recently declared independence. In one of his anti-Revolution caricatures, Gillray chose the captions 'French Liberty' and 'British Slavery' to contrast a dark-skinned emaciated Frenchman with a pale well-fed Briton.[5]

Although Newton's split figure is exaggerated and biased, like other caricatures this 'real san-culotte' provides valuable snapshot evidence of British fears at this particular time. After all, people then were not privy to our retrospective knowledge of what happened next: nobody knew whether the chaos in Paris and Saint Domingue would spread to Britain. Compounding British anxieties, in India at this time the anti-British Tipu Sultan was demolishing churches and incarcerating Catholics, violent acts often interpreted as an Islamic campaign to eliminate Christianity and subvert European authority. For British conspiracy theorists, it was only too easy to imagine that their treasured way of life was under attack from a conglomerate of threatening outsiders—French Jacobins, Napoleonic imperialists, Muslim aggressors, disobedient African slaves.

Anti-Jacobin reactionaries also perceived threats emanating from within their own country—workers who demanded the voting rights available only to privileged men, women who insisted on better education, abolitionists who urged that slavery be ended. One tactic adopted for emphasizing these threats was to conflate them. Enlightenment reformers deployed similar rhetorical devices in their own propaganda, choosing the language of enslavement to describe not only Africans being bought and sold for their labour, but also women who were subservient to their husbands, and men who were denied the vote. Capel Lofft, a radical dissenting lawyer and anti-Pittite, declared that 'to be enslaved is to have no will of your own in the choice of law-makers; but to be governed by rulers whom other men have set over us'.[6] Pointing to the new Republic across the Atlantic, William Blake portrayed British women as oppressed by the chains of marriage:

> Enslav'd, the Daughters of Albion weep; a trembling lamentation
> Upon their mountains; in their valleys, sighs towards America...[7]

Women, slaves, and indigenous peoples had long been interchangeable as oppressed minorities: 'in all the Eastern parts of the World,' an English essayist (Judith Drake) complained towards the end of the seventeenth century, 'the Women, like our Negroes in our Western Plantations, are born slaves, and live Prisoners all their Lives.'[8] The comparisons still seemed valid a hundred years later. In Britain, voting was restricted to male property owners, women belonged either to their fathers or their husbands, child labour was prevalent, only rich Anglican men could go to university, and the economy depended on the slave-fuelled trade between Africa, Europe, and the Americas.

The French and American revolutionaries promised to eliminate the privileges of birth, but they focused on class and paid little attention to colour or gender. Even female campaigners such as Mary Wollstonecraft, who had lived in Paris during the Revolution, accepted that women should be subservient to men, while France showed little intention of withdrawing from the lucrative slave trade. When America's founding fathers were drawing up the new constitution, slavery proved such a contentious issue that their only decision was to postpone any decision for the next twenty years. Until 1808, they declared, helping slaves to escape would be illegal—and in censuses, each slave would count as only three-fifths of a person. In other words, the slave trade was federally endorsed in this new land of equality. Samuel Johnson highlighted the paradox of American colonialists who demanded freedom yet practised slavery: 'How is it that we hear the loudest yelps for liberty among the drivers of negroes?' he demanded.[9]

Nowadays, not all Europeans and North Americans are equally well off, but they do have equal rights. Or do they? Slavery was made illegal during the nineteenth century, and in 1971, Switzerland became the last European country to introduce female suffrage. Even so, Britain has never let the revolutionary ideals of 'Liberty, Equality & Fraternity' completely overturn national tradition. The throne and hereditary peerages are still acquired by birth rather than merit—and their rules still give priority to men over women.

I suspect that Darwin would have seen nothing wrong with this blurry version of equality. However threatening he might have seemed to the *Anti-Jacobin*, he was an armchair reformer rather than a rabid revolutionary. For him, the key to a better Britain was knowledge. He wanted to improve education for boys and girls of all classes, to reduce illness by educating patients about health care, to show farmers how agricultural yields could be increased, to help factory owners boost industrial output by using the latest inventions. Although it might seem hard to imagine that anybody could object to such fine ambitions, those in the upper echelons of society had much to lose.

Darwin's friend Joseph Priestley triumphantly spelt out the dangers for the ruling classes. They should, the chemist warned, tremble at the prospect of electrical machines and air pumps, whose use depended on intelligence rather than inheritance. They were not only instruments of science, but also agents of democracy—although aristocrats were the traditional rulers of society, experimenters would control nature. Recognizing that knowledge brings power, another of Darwin's colleagues, the industrialist Matthew Boulton, supported James Watt while he was developing massive steam engines that promised to boost efficiency and profits. When Boulton escorted Samuel Johnson's biographer James Boswell around his Birmingham metalware factory, he gestured expansively across his machines and the seven hundred workers who tended them. 'I sell here, sir,' he proclaimed, 'what all the world desires to have—POWER.'[10]

Boulton was only a buckle-maker when Darwin first met him, but they soon became close friends. Both ambitious men, their interests meshed: the relatively uneducated metalworker gained entry into intellectual company, while the self-taught inventor forged links with rising industrial entrepreneurs. Such a collaborative inter-class relationship, in which neither partner acted as patron, would have been near unthinkable further south or fifty years earlier. Gradually, similarly diverse acquaintances intermingled and coalesced to create an informal Midlands network known as the Lunar Society. Although neither as large nor as scientifically focused as London's Royal Society, this group demonstrated that technological nous was sufficiently powerful to transform the world.

CHAPTER SEVEN

The Lunar Society

But far above and far as sight endures
Like whips of anger
With lightning's danger
There runs the quick perspective of the future.

Stephen Spender, 'The Pylons' (1933)

Without the Lunar Society, Russia rather than Britain might have become celebrated as the home of James Watt and the Industrial Revolution. In 1775 the Russian Imperial Government offered Watt an impressive salary to bring over his expertise on steam engines. As one of his closest friends, Darwin was horrified at such a brain-drain possibility. 'Lord, how frighten'd I was, when I heard a Russian Bear had laid hold of you with his great Paw,' he wrote to his Lunar colleague. Getting carried away with his own imagery, Darwin assumed that Watt would (unlike me) know about Cacus, the mythical fire-eating giant who dragged cattle into his cave by their tails. 'Russia is like the Den of Cacus,' he warned Watt; 'you see the Footsteps of many beasts going thither but of few returning. I hope your Fire-Machines will keep you here.'[1]

Watt yielded, staying behind to produce steam engines in Birmingham and become a core member of the Lunar Society, the loose association of forward-thinking colleagues often credited with turning Britain into Europe's first industrial nation—men such as the potter Josiah Wedgwood, the chemist Joseph Priestley and the factory owner Matthew Boulton. Membership fluctuated, but from the mid-1760s to the early 1790s around fourteen men met together each month on the nearest Monday to the full moon, when their routes home through the Midlands countryside would be well lit. The name they chose for themselves derives from this arrangement (think *lundi* and *la lune*, French for 'Monday' and 'the moon'), but was presumably also a self-mocking joke (think 'lunatic').

Darwin liked to regard himself as a 'mechanical philosopher', although 'inveterate tinkerer' would be a more accurate, if less flattering, description.[12] In contrast, several of the other Lunar men were hard-headed industrialists determined to make money and climb socially. By hymning their inventions in his poetry, Darwin acted as an unpaid PR agent for individual members as well as for industrial innovation more generally. Similarly, Joseph Wright of Derby—Darwin's patient and friend—became the group's unofficial artist; although not a member, Wright advertised their initiatives by painting them, their inventions, and their factories.

Priestley formulated the nearest thing to a Lunar manifesto. 'We had nothing to do with the *religious* or *political* principles of each other,' he declared; 'We were united by a common love of *science*, which we thought sufficient to bring together persons of all distinctions, Christians, Jews, Mohametans, and Heathens, Monarchists and Republicans.'[13] These Midlands men shared the overriding goal of progress, but they had no government backing, no metropolitan head-office, and saw no need for bullet-point lists of aims and objectives. Science, technology, and industry dominate present-day existence, and from the mid-twentieth century on, historians have paid them increasing attention. The Lunar friends are now celebrated as crucial founding fathers of modern Britain, and in this alphabetical list I have (somewhat anachronistically) indicated their scientific activities, using a * to mark the eleven who were also Fellows of the Royal Society:

    *Matthew Boulton (metal manufacturer)
    *Erasmus Darwin (physician, botanist)
     Thomas Day
    *Richard Edgeworth (inventor)
    *Samuel Galton (gunmaker)
    *Robert Johnson
    *James Keir (chemical manufacturer)

*Joseph Priestley (chemist)
William Small (physician)
Jonathan Stokes (physician, botanist)
*James Watt (engineer)
*Josiah Wedgwood (mineral expert, inventor)
*John Whitehurst (geologist)
*William Withering (physician, botanist)

But the past is always open to reinterpretation. As I learnt more about this disparate group, attempting to view them as individuals and from the perspective of their contemporaries rather than mine, I became increasingly interested in their bids for social reform. Following the pronouncements of Francis Bacon—Elizabethan philosopher and heroic figurehead of the Royal Society—they believed that scientific advances could and should benefit society. Even so, some of these men knew little about steam engines, pottery or medicine, and as a consequence have received relatively little attention, although prominent at the time.

Through thinking about Darwin, I arrived at a rather different vision of the Lunar men, one that considers their political activities as well as their scientific ones. This second list shows the same men, in the same order, but this time highlighting their interest in reform or social improvement. Again, eleven are marked*, but the collective character of the group appears very different.

Matthew Boulton
*Erasmus Darwin (poet, political radical)
*Thomas Day (social reformer, abolitionist poet)
*Richard Edgeworth (educational reformer)
*Samuel Galton (Quaker)
*Robert Johnson (Anglican minister)
*James Keir (supported French Revolution)
*Joseph Priestley (Dissenting minister, political radical)
*William Small (tutor of Thomas Jefferson, supporter of American Revolution)
Jonathan Stokes
James Watt
*Josiah Wedgwood (abolitionist)
*John Whitehurst (supported American Revolution)
*William Withering (popular educator, especially of women)

For these ambitious energetic men, social progress went hand-in-hand with scientific and technological improvement. By implementing innovation, they hoped not only to increase their own profits but also to make Britain a better place—by which they understood not only more prosperous but also more equal.

Lunar Friends

Frustratingly, there are no Lunar minutes for historians to pore over, so their discussions have to be inferred from the correspondence that survives. Their letters reveal how intimately these men's lives were interleaved, and what close friends they became over the years as the group expanded. This sense of camaraderie comes across vividly in a letter Darwin wrote to Boulton after being forced to miss a monthly get-together. 'I am sorry the infernal Divinities, who visit mankind with diseases, and are therefore at perpetual war with Doctors, should have prevented my seeing all you great Men at Soho today,' he wrote regretfully. 'Lord! what inventions, what wit, what rhetoric, metaphysical, mechanical, and pyrotecnical [*sic*], will be on the wing, bandy'd like a shuttlecock from one to another of your troop of philosophers! while poor I, I by myself I, imprizon'd in a post chaise, am joggled, and jostled, and bump'd and bruised along the King's high road, to make war upon a pox or a fever!'[14]

The Lunar men can appear overwhelmingly dedicated to science because most of them were—like Darwin—Fellows of the Royal Society (FRS). However, they were not scientists in the modern sense. The oldest, John Whitehurst, illustrates that modern categories can be unhelpful for describing their activities. As an eminent Derby clockmaker, Whitehurst was a traditional craftsman who adapted his skills to design precision instruments: invaluable work, but not cutting-edge research. In retrospect, his ideas about rocks presage scientific theories of evolutionary change, but the academic discipline of geology did not yet exist, and he was troubled by the contradictions between what he observed in the local countryside and the lessons he had learnt from the Bible. Long before Charles Darwin was born, Whitehurst challenged conventional thought. If God had created the cosmos exactly as it is now, he asked himself, then why was there so much evidence of change?

Another example of an FRS with little formal scientific training is Boulton, one of the most successful Lunar entrepreneurs, who expanded his father's small metalworking business by introducing steam engines, minting copper coins and producing fine silver ornaments for the luxury market. His success depended on investigations that sound scientific but which stemmed from industrial requirements and took place in a factory, not a laboratory: finding durable substances for heavy machinery, testing unusual minerals for decorating tableware, assessing alloys for strength and weight.

Similarly, although Wedgwood never went to university he developed a high-temperature thermometer (pyrometer) for his pottery kilns that proved invaluable for chemical experiments. To boost sales of his vases and plates, Wedgwood embarked on an intensive, systematic, and well-documented series of experiments; however, unlike a professional mineralogist, he received no salary for his research. 'I scarcely know without a good deal of recollection whether I am landed gentleman, an engineer or a potter,' he wrote to Boulton; 'for indeed I am all three by turns...'[15]

In contrast, the factory owner James Keir was a well-educated military man whom Darwin had first met when they were both university students at Edinburgh. Although there were then no degrees in industrial chemistry, Keir became rich and famous through inventing mass-produced soap. Increased hygiene was one of the few real contributors to improved health at the end of the eighteenth century, and cheap soap (along with cotton clothes and the abolition of window tax) was far more influential than most medical innovations. At the same time as making his fortune, Keir was also fulfilling the Enlightenment ideal of improving the nation.

The Lunar Society is remembered for its industrial legacy, but its members also bequeathed a more egalitarian society. '[W]e live but to improve,' wrote James Watt's wife Annie to their son.[16] To make people's lives better, the Lunar men aspired not only to introduce technological innovations, but also to reform social structures. 'Wedgwood' now has brand identification as expensive blue and white pottery, but Josiah Wedgwood was a leading campaigner in the movement to abolish slavery. Priestley is celebrated as a great chemist, but he was notorious for his unorthodox religious views and his radical political tracts. Like both of them, Darwin was an FRS, but he also condemned slavery and strongly supported the revolutionary ideals of his friend Benjamin Franklin.

The newly developing industrial north provided unprecedented opportunities for workers to become employers, and the mixed composition of the Lunar Society reflected their belief that wealth should be based on ability rather than on birth and family pedigree. For instance, Darwin's son Robert came from a privileged background, but he married Wedgwood's daughter Susanna. Only one generation away from uneducated poverty, she inherited a fortune worth well over a million pounds in modern terms. Such a cross-class alliance would then have been unthinkable in conventional circles, as one of the sons of Robert and Susanna discovered when he moved south and was spurned for his lowly origins.

The Lunar Society's reforming ideals were given a fresh twist by Richard Lovell Edgeworth and Thomas Day, two younger men who were both enthused by the latest French ideas about education. An inspired inventor from Ireland, Edgeworth immediately impressed Darwin with his mechanical ingenuity. Day was different, a philosopher more interested in metaphysics than physics. Darwin initially found him uninteresting (perhaps the self-absorbed guest was insufficiently polite about his host's inventions), but once he had learnt to appreciate the intellect of this unkempt but wealthy young man, he changed his mind.

In the botanic garden that Darwin cultivated to impress Elizabeth Pole, he moulded female nature to his own tastes. Similarly, Day decided to fashion a wife for himself. Lifting two Lolita-age girls out of their orphanages, Day brought them up in France according to the educational philosophy of Jean-Jacques Rousseau, teaching them to love reading rather than clothes and to live naturally without the trappings of modern civilization. He soon discovered that encouraging his protégées to develop minds of their own had a foreseeable but undesirable consequence—they refused to do what he wanted. Apprenticing one to a milliner, Day took the other to live near Darwin at Lichfield, where he strengthened her fortitude by sprinkling her with hot sealing wax and firing pistols at her skirt. After that failed to work, he sent her off to boarding school, convinced that he could never convert her into a worthy wife and mother for his children. In Anna Seward's acerbic words, Day's 'trust in the power of education faltered' and he abandoned these experiments. (He was later kicked to death by a colt he was trying to train with kindness rather than punishment.)[17]

Day was a less ludicrous character than my account of this educational experiment makes him seem. His moralizing children's book, *Sandford and Merton* (1783–9), remained a best-seller for eighty years. It describes how, under the guidance of a patient teacher, a spoilt rich brat comes to recognize his inferiority to his impoverished but upright playmate. A fervent supporter of the American revolutionaries against the British government, Day was also a leading anti-slavery campaigner whose heart-rending poems did much to inflame public outrage at the conditions on transatlantic slave ships. The first, 'The Dying Negro' (1773), was based on a true story, and takes the form of a love letter from an escaped African slave who has chosen death rather than transportation. A newly converted Christian, he is writing to his intended wife, a white servant in the same household. In the closing lines, the slave-author prays as he prepares to commit suicide:

> O lead me to that spot, that sacred shore,
> Where souls are free, and men oppress no more![18]

In a similarly empathetic vein, Darwin poetically evoked the feelings of African mothers as they watch their children being shipped abroad, imagining how the American plant *Cassia* must grieve as its black seeds are washed across the Atlantic to Norway.[19]

Despite Priestley's utopian portrayal of a collaborative community, internal conflicts did divide the group, especially over the American and French Revolutions. In particular, Boulton and Keir were less radical in their political outlook than Priestley and Darwin. Darwin's personal sparring partner was William Withering, the doctor who had converted a traditional country remedy into a pharmaceutical drug, digitalis. Both provincial physicians, both prickly, Darwin and Withering continually jibed at each other in public as well as in private, exchanging bitter letters about priority and medical treatments. In popularizing Linnaean botany, they competed for the same female market, but Withering believed in protecting women from the sexual implications of the new classification system. The recipients were not always grateful—one of his correspondence pupils protested that 'the Generalty of men have Agreed that Women ought to be kept in perpetual Ignorance & the most profound Darkness, respecting every part of Literature beyond a Book of Cookery'.[20]

That this network of independent pioneers managed to stay together for so long was due partly to the diplomacy of William Small, a Scottish doctor who had been working in America and returned with a letter of introduction from Benjamin Franklin. Like Watt, Small turned down an invitation to practise in Russia, preferring to settle in Birmingham, where he maintained an orthodox medical career rather than dedicating himself to the more controversial domains of chemistry and machinery. Although not a technical innovator, his expertise in tactful soothing proved valuable. It was from Small that Darwin acquired the strategy of silencing his critics by drinking tea with them, a tip that he passed on to his son Robert—'it is best to shew a little attention to those who dislike one, at public assemblies; and it generally conciliates them.'[21]

Profiting from improvement

By offering each other technical advice, commercial contacts, and emotional support, the Lunar colleagues made personal fortunes but also cooperated to transform British science, industry and agriculture. By building canals, they literally transformed England, criss-crossing the land with waterways that linked the east to the west and the north to the south. As labourers started migrating from fields to factories, from villages to cities, the urban population expanded and provincial centres began to rival the metropolis. Raw materials—clay and coal, iron and wood—could now be efficiently shipped across the country to distribution hubs, and then sent on to London or the west-coast ports of Liverpool and Bristol.

This rapid improvement in internal transport was funded not by the government but by private financiers—men like Wedgwood who were inspired by personal ambition as well as grandiose visions of the future. Most of the Lunar members invested their own wealth in these venture schemes, and several became involved at a practical level. For instance, Darwin organized a side-cutting to power a local mill, Small tactfully negotiated disputes over water supplies and land purchases, while Boulton ran a lucrative business manufacturing metal components for barges and locks.

The Lunar men also engaged in less momentous collective projects. With the original aim of helping Wedgwood, Darwin spent years developing a windmill for grinding minerals. This gradually became a joint venture—Whitehurst suggested improvements, Watt came over from Birmingham to give technical advice, and Edgeworth pitched in to build a scale model. For his part, Wedgwood turned for advice to Priestley's chemical textbook when he was designing some pottery equipment—and after Priestley and Watt advertised the result, he sold it commercially. Pooling their expertise, Keir and Wedgwood experimented together on glass and clays—and for geological information, they relied on Whitehurst and Boulton.

As well as technical help and financial backing, the Lunar friends gave each other psychological comfort, especially during the frequent illnesses and deaths of family members. Trying to console Watt during one of his recurrent fits of depression, Darwin mixed jocularity with medical advice: 'Why will you not live at Derby? I want learning from you of various kinds, and would give you in exchange chearfulness [sic]. . . . Why the d—l do you talk of your mental faculties decaying, have not you more mechanical invention, accuracy, and execution than any other person alive? . . . Your headaches and asthma would recieve [sic] permanent relief from warm bathing I dare say—but perhaps you are too indolent to try its use?'[22]

Buoyed up by his success with *The Loves of the Plants*, Darwin co-opted his Lunar friends to help him produce its sequel, *The Economy of Vegetation* (published as its prequel in *The Botanic Garden*). Pointing out that he was offering them free publicity, Darwin set up his second major poem as a collaborative Lunar project, a manifesto advertising national industrial progress. Watt contributed details of steam engines, Boulton provided information on his copper coining processes, and Edgeworth supplied helpful criticisms (although Darwin's self-defensive reply suggests these were less welcome).[23]

For his Lunar colleagues, Darwin presented his literary initiative as a commercial undertaking: perhaps he felt the need to justify his sedentary domestic occupation that differed markedly from their practical industrial schemes. Boasting to Watt, Darwin reported that 'the Loves of the Plants pays me well'. That was true—apparently he sold the copyright for an astonishing £800. But I was not so sure about his next claim. '[A]s I write for pay, not for fame,' he declared, 'I intend to publish the Economy of Vegetation...'. My suspicions were heightened when I discovered another letter from Darwin sent six months later. Addressed to a close friend of the prime minister, it tentatively enquired whether he might be considered as the next Poet Laureate, a post that would bring in a comparatively low annual salary of £100. 'The Fame of it is no object to me,' he insisted. I remain unconvinced.

The Lunar men saw nothing wrong in promoting their own financial interests: on the contrary, they claimed that improving their personal positions would benefit the entire nation. This emphasis on the rewards of individual gain typified the new Enlightenment code of morality. In the past, being virtuous had entailed suppressing any wish for advantage, and instead letting others come first. According to the analysis introduced by the Scottish philosopher Adam Smith, full recognition should be given to the realities of human nature: for most people, the strongest incentive for action is to better their own situation. Smith argued that this was not necessarily a selfish way of behaving, but one that benefited manufacturers and consumers alike. For producers to make financial profits, they must sell goods—but these are desired by purchasers to improve their lives. In this win-win situation of enlightened self-interest, the economy booms, there are more things to buy, and more money to buy them with. The Enlightenment invented conspicuous consumption as well as steam engines and rationality.

Despite their high-sounding ideals and hopes for the future, the group's reforming zeal was blunted by personal rivalries and conflicting goals. Enthused by profits and progress, the Lunar industrialists built ever faster and more powerful machinery, but not everybody gained from these changes. Their ambitious entrepreneurship involved costs, compromises, and concealments. Hidden between the lines of Darwin's poetry lies evidence of collective theft—the freedom and the labour that Lunar factory owners took from their workers, whose existence is scarcely even mentioned.[24]

When Boulton developed his Soho estate outside Birmingham, he aimed to improve working conditions by providing well-lit rooms, health care, and education—but clearing the site entailed demolishing its existing cottages and moving its inhabitants. Although Darwin rejoiced that Boulton could mass-produce thirty thousand uniform coins an hour, he showed no concern for the welfare of either the displaced hand-workers or the four young boys trying to keep up with the machinery's rapid output.[25] When he praised Wedgwood for making teapots that were of far higher value than the minerals they were created from, Darwin neglected to mention the risks to workers and consumers arising from repeated failures to make a lead-free glaze.[26]

Since Wedgwood had grown up in poverty, and was one of Britain's leading abolitionists, you might have expected him to treat his workers kindly. On the contrary, he was a strict disciplinarian who boasted that 'my name has been made such a scarecrow to them, that the poor fellows are frighten'd out of their wits when they hear of Mr W coming to town, & I perceive upon our first meeting they look as if they saw the D[evi]l'. When his staff rioted, he gave them a pep talk on the benefits of industrialization. In his hyper-efficient factory, employees clocked in and out, trooped back to work at the summons of a bell, and paid fines for breaches of behaviour.

Wedgwood wanted, he declared, to 'make such *machines* of the *Men* as cannot err'. Following Adam Smith's guidance on the division of labour, he increased productivity but decreased morale by obliging his workers to perform simple repetitive actions instead of being responsible for a complete piece. He showed few qualms about firing people he no longer needed. 'Gold, the most precious of all metals is absolutely kicked out of doors,' he wrote when fashions changed; ruthlessly, he spelt out the consequences: '& our poor Gilders I believe must follow it'. Women worked for Wedgwood not because he wanted to help them, but because they were so desperate for jobs that he could get away with paying them lower wages than the men.[27]

As another example of concealed labour, the flirtatious flowers in *The Botanic Garden* are very different from the capable women who made it possible for the Lunar Society to function. Because the meetings took

place in private homes rather than in the male territory of clubs and coffee houses, they were domestic events at which women and children
were visible. Wives and daughters were regularly recruited to keep
accounts, wash out instruments, and record results, and several were
intimately involved in the scientific research that made their men
famous. Wedgwood's wife Sally advised him on pottery patterns and
kept up the shorthand notes in his logbook—and soon after their daughter Susanna was born, Wedgwood despatched Sally's spinning wheel to
the junk room so that she would have more time to work for him. Priestley's wife Mary was responsible for keeping his experimental mice warm
on the mantelpiece, but was disconcerted to find chemical flasks and
minerals stowed away in her travelling trunk of neatly packed clothes.
Watt sent technical letters about chemistry to his wife Annie, whose
family was in the bleaching trade.[28]

Politics turned out to be the divisive issue that finally rent the group
apart. Like Keir and Withering, Darwin was enthusiastic about radical
reform. 'Do you not congratulate your grand-children on the dawn of
universal liberty?' he asked Watt; 'I feel myself becoming all french [*sic*]
both in chemistry and politics.'[29] While Watt grumbled about the lack of
chemical information filtering out of battle-torn Paris, Priestley became
such an ardent supporter of the French Revolution that his house was
burnt down during a Birmingham riot in 1791. Even though his Lunar
colleagues loyally supported him, they were worried about their own
safety. The Terror was escalating in Paris, and British establishment propaganda (in the *Anti-Jacobin* and elsewhere) was directed against French
chemistry and experimental research—precisely the sort of activity in
which many of them were engaged.

Priestley eventually fled to America, and the Lunar Society faded away
as the members grew older and concentrated on their own interests. Edgeworth was remote in Ireland, Boulton was plagued by kidney problems,
and Withering lamented that 'our Lunar meeting...has never flourished
since the departure of Priestley. The Members are to a Man either too
busy, too idle, or too much indisposed to do anything; and the interest
wch everyone feels in the state of public affairs draws the conversation out
of its proper course.'[30]

When Priestley was being attacked by Birmingham rioters, Darwin had been living in Derbyshire with Elizabeth Pole for ten years, and was caught up in arguments about the engravings for his next long poem, *The Economy of Vegetation*. In particular, he wanted to promote the Portland vase, an exceptionally fine piece of decorative Roman glass that was many centuries old, but named after its Enlightenment owner. Wedgwood spent four years trying to reproduce its graceful white figures set against deep blue glass, and when he finally succeeded, public fascination was so high that the entry tickets for the private viewing had to be restricted. Darwin was determined to find a first-class illustrator to celebrate Wedgwood's achievement, and it was only after much wrangling that he agreed to take his publisher's advice and accept an engraver of whom he'd never heard. This unknown artisan was called William Blake.

Several stunning pictures by Blake and Fuseli were originally bound in amongst the verses of Darwin's *Economy of Vegetation*. However, when I looked at electronic versions and recent reprints, only the words were there. I realized that to understand readers' reactions at the time, I needed to look at the original version—and also to read it from cover to cover. Instead of relying on other historians' opinions, I had to start at the beginning and keep going until I got to the end. My personal impressions of what I found make up the next chapter.

By reporting my first-hand experience, I hope to explain why Darwin thought it made sense to weave together such superficially incompatible topics as paradise and porcelain, Jupiter and Miss Jones, slaves and sea horses. If you feel that your naive guide—me—has provided a questionable interpretation, then there is only one thing to do: read the poem for yourself.

CHAPTER EIGHT

The Economy of Vegetation

The day has been, I grieve to say in many places it is not yet past, in which the greater part of the species, under the denomination of slaves, have been treated by the law exactly upon the same footing as, in England for example, the inferior races of animals are still.

Jeremy Bentham, *An introduction to the principles of morals and legislation* (1789)

At first, even the title seemed strange. To me, the words 'economy of vegetation' suggested cost-efficient farming, and had little to do with either factories or botany. And when I remembered John Armstrong's *The Oeconomy of Love* (1736), I was still more perplexed. Repeatedly reissued in cheap editions, this slim volume would have been familiar to many of Darwin's readers. Both *Economies*—of vegetation and of love—were doctors' attempts to explain science in verse, although Armstrong's poem is far more sexually explicit than Darwin's.[31]

I later discovered I was not alone in feeing confused. When James Watt received a letter from Darwin asking for detailed information about steam engines, he sounded puzzled. 'I know not how steam-engines come among the plants,' he replied; 'I cannot find them in the Systema Naturae, by which I should conclude that they are neither plants, animals, nor fossils.'[32] Watt's reference to *Systema Naturae*, the standard manual of natural classification, led me to recognize that industrial and botanical plants might belong together. Its author, Carl Linnaeus, was also responsible for *The Economy of Nature*, perhaps the inspiration for Darwin's own title— *The Economy of Vegetation*.[33]

For Darwin and Linnaeus, 'economy' referred not so much to managing money, but more to understanding how systems are structured and organized. This was a time when 'industry' meant diligent hard work and 'economy' was closer to the modern 'ecosystem' than to international finance: like 'ecology' (invented in the nineteenth century by Ernst Haeckel, the German Darwinian), its roots lie in the Greek word for 'household'. God's design could be revealed by classifying plants, and His world could be improved by transporting crops around the globe and cultivating them in different environments. Inspired by his faith in divine economies, Linnaeus became convinced that expensive imports could be minimized by persuading coffee, tea, and other luxury crops to grow in Sweden (no, this was not a successful project).[34]

Biological and economic parallels survive in modern expressions: thrifty farmers husband their fields as well as their savings, and economies recover as though they were sick patients whose healthy equilibrium has been restored. The natural world was envisaged as a self-regulating hierarchy in which each animal or plant forms its own small, balanced economy that interacts with larger and smaller ones in mutual interdependence.[35] Darwin explained how leaves maintain 'the vegetable œconomy', nourishing their parent plant by acting like the lungs of animals and fishes; similarly, plants themselves contribute towards the survival of the animals that feed off them.[36]

During the eighteenth century, the keys to making economies thrive were order and circulation. Subjects revolved around their monarch like planets around the sun or children around their father, while human health and stability were maintained by fluids flowing freely round the body; similarly, the medicines, money, and raw materials essential for the nation's well-being circulated through a well-regulated economic structure from which all participants profited. Given this ordered, circulatory model, technological progress seemed to benefit everyone. Using machines to increase productivity would give workers more time for prayer and education, so that the moral and financial state of individuals, employers, and the nation would improve together. For Darwin and his eighteenth-century readers, there were close ties between the political and the cosmic, the biological and the ethical, the educational and the financial aspects of existence.

It was time to move on from the title page and start the poem itself. Before I got there, I encountered a short 'Apology', which purported to explain the verses lying ahead. Lacking the education of an Enlightenment gentleman, I was bewildered by Darwin's references to Egyptian hieroglyphs, classical myths, and Rosicrucian doctrines. What could they have to do with either plants or technology—or, indeed, with each other? In the hope that understanding would arrive later (which it did—I eventually went back to this topic when I was thinking about *The Temple of Nature*), I put that problem to one side and embarked on the poem's four cantos, which correspond to the four Aristotelian elements of fire, earth, water, and air.

Canto I: Fire

Following a practice common in eighteenth-century poetry, at the beginning of each canto Darwin provides an Argument, a telegram-like summary resembling a contents page. At a glance, you can see that he interprets 'fire' broadly: this section starts with meteors, and then perambulates through the aurora borealis, volcanoes, glow-worms, electricity, and the constellations. In between those fiery topics are scattered some unexpected references, including Venus's encounter with the Cyclops, Memnon's Harp, and Abyla and Calpe. Although presumably familiar to Darwin's readers, they were not to me.

Somewhat apprehensively, I started reading. I immediately felt heartened by recognizing the first two lines, which I had seen reproduced in a footnote to 'The Loves of the Triangles':

> STAY YOUR RUDE STEPS! whose throbbing breasts infold
> The legion-fiends of Glory, or of Gold![37]

Although this couplet meant little to me, it clearly did to the *Anti-Jacobin* satirists, who expanded it for the beginning of their own poem:

> STAY your rude steps, or e'er your feet invade
> The Muses' haunts, ye Sons of War and Trade!
> Nor you, ye legion Fiends of Church and Law,
> Pollute these pages with unhallow'd paw![38]

By starting in this way, the parodists made it clear that their poem was no trivial entertainment, but a political manifesto directed against anyone who threatened the hallowed British virtues of religion and stability—in other words, Jacobins and their allies. Despite its frivolous title, their satire was intended to be not merely an amusing frippery about soft-porn botany, but instead a calculated attack on radicalism that drew on all of Darwin's publications up till then.

Returning to Darwin's *Economy of Vegetation*, I noticed that many stanzas opened with inverted commas. But who was saying what to whom? I worked out that the first speaker is the presiding Genius of Darwin's ornamental garden near Lichfield, who is preparing this rural love nest for a visit from the female Goddess of Botany. Once the guest has arrived, in suitably splendid style (an aerial descent later borrowed by Shelley for *Queen Mab*), she gathers the nymphs of fire around her for an extended soliloquy that starts with the origins of the universe, reviews (verbosely) its igneous development, and ends with Elijah on Mount Carmel.

Throughout, Darwin smuggles in large doses of technical science. His footnotes are even longer than in *The Loves of the Plants*, often squeezing out all the verses from a page, while the essays at the end take up as many pages as the poem itself. For example, the short section from line 97 to line 105 is expanded by three long sections of prose. In the first, Darwin embroiders 'Nymphs of primeval fire' with an academic disquisition on the nature of heat (antagonizing *Anti-Jacobin* adherents by using the French term 'Calorique'); four lines later he provides an extended riff on evolution, including the progressive formation of the Earth's crust, redundant organs, and Nature's striving towards perfection through constant change; and four lines after that, he describes a recent paper in the *Philosophical Transactions* supporting his assumption that at the creation, stars shot out through space like grains of gunpowder.

In my opinion (but then, I'm not an eighteenth-century enthusiast), Darwin contrives links that are tenuous to the point of being laughable. He manages to fit in the death of Professor Richman (who should have been more careful when experimenting with lightning), the great Egg of Night (an ancient Greek myth about creation and love), and an attack on a whale (harpooned by sailors as it searched for warmer waters). Ingeniously, without deviating too far from his overall theme, Darwin manages to jump from Don Quixote (the levelling value of gunpowder) to Hercules (as powerful as an aeroplane), to paralysis (nerves detect heat) and sympathetic inks (visible only when warmed).

This canto celebrates human achievement. In particular, Darwin lavishly plugs two of his Lunar colleagues, Watt and Boulton, transforming into heroic couplets the technical details he had asked them to send him. Within fifty years, Darwin forecasts optimistically, travel and industry will have been transformed by steam engines:

> Press'd by the ponderous air the Piston falls
> Resistless, sliding through it's [*sic*] iron walls;
> Quick moves the balanced beam, of giant-birth,
> Wields his large limbs, and nodding shakes the earth.[39]

As well as praising innovation, this poem is also a nationalistic paean to British achievement. Even the *Anti-Jacobin* could hardly protest about this resounding climax to Darwin's description of Boulton's coin-pressing machine:

> Hard dyes of steel the cupreous circles cramp,
> And with quick fall his massy hammers stamp.
> The Harp, the Lily and the Lion join,
> And GEORGE and BRITAIN guard the sterling coin.[40]

But the reviews make it clear that many people did object strongly to this particular couplet, which stems directly from the opening of Lucretius' *On the Nature of Things*:

> When LOVE DIVINE, with brooding wings unfurl'd,
> Call'd from the rude abyss the living world.[41]

Darwin's readers would have recognized its similarity to the Lucretian equivalent, which reads (in a modern translation): 'Mother of Aeneas ... life-giving Venus, it is your doing that under the wheeling constellations of the sky all nature teems with life ... into the breasts of one and all you instill alluring love, so that with passionate belonging they reproduce their several breeds'.[42] According to conventional natural theology, God the Designer carefully planned everything in His creation well in advance. In contrast, in a Lucretian universe, matter and life arise through chance encounters. Such a philosophy was anathema not only to the authors and readers of the *Anti-Jacobin*, but to most British people of this period. Although Darwin does mention God several times, he gives Him the deistic role of remote caretaker, and instead makes sexual energy the fundamental driving force of the cosmos.

Canto II: Earth

Turning to glance at the next canto, my first thought was 'Oh no, more of the same' (research can be tedious as well as thrilling). As before, I was taken aback by the diversity of Darwin's topics on the general theme of earth. Whereas I could understand the relevance of marble and magnets, of coal and copper, of mines and manures, I felt alienated by the inclusion of characters from Greek myths—Venus and Vulcan, Adonis and Apollo. I did manage to resolve some puzzling references: Hannibal crossing the Alps turns out to be a scientific justification of the story that his army bored its way through the mountains with fire and vinegar, while St Peter being delivered from prison is literary imagery for plant roots forcing upwards through 'obdurate clay' to reach the light (the Quaker prison reformer John Howard gets a brief mention here as well).

More obviously, I could admire Darwin's ingenuity in fabricating advertisements for his Lunar colleagues by focusing on geology, mining, pottery, and metal manufacture. He credits Whitehurst with his geological observations, cites Priestley several times for his expertise on gases, and even manages to weave in a reference to Lichfield, apparently the birthplace of the first person to make an orrery (the rotating model of the planets painted so evocatively by Wright of Derby, the Lunar Society's unofficial artist). Darwin reserves his longest encomium for his close friend Wedgwood. In a potted version of pottery's history, he contrives to reach a climax in Etruria, the region in central Italy then renowned for fine antique vases: perfect cue for modern-day 'Etruria', the name of Wedgwood's factory in Staffordshire. As ever, Darwin packs his dense notes with erudite speculations, while churning out as if effortlessly his rhyming praise of Wedgwood's skills:

> Charm'd by your touch, the kneaded clay refines,
> The biscuit hardens, the enamel shines...[43]

Despite the similarities of format between the first two cantos, this second one is more contentious. Dark and dissonant, it is written in abrasive hissing vocabulary and shot through with political imagery. At first, the mood is calm, as the Goddess of Botany addresses the gnomes who are welcoming the birth of a globe resembling paradise. This initial tranquillity is disrupted when volcanoes rip apart the Earth to create the Moon, confirming Darwin's view that the universe is a site of upheaval rather than stability, a troubled war zone where perpetual battle is waged between the forces of good and evil:

> GNOMES! how you shriek'd! when through the troubled air
> Roar'd the fierce din of elemental war;
> When rose the continents, and sunk the main,
> And Earth's huge sphere exploding burst in twain.—[44]

About halfway through this second canto, Darwin shifts into a political gear. His friend Benjamin Franklin provides a convenient opportunity for sliding between an encomium to his experiments on thunderous lightning and the political demons being purged from America:

> While from his eyry shriek'd the famish'd brood,
> Clenched their sharp claws, and champ'd their beaks for blood,
> Immortal FRANKLIN watch'd the callow crew,
> And stabb'd the struggling Vampires, ere they flew.
> —The patriot-flame with quick contagion ran,
> Hill lighted hill, and man electrised man.[45]

Elaborating this theme, Darwin next imagines a warrior of Liberty crossing the Atlantic:

> The Warrior, LIBERTY, with bending sails
> Helm'd his bold course to fair HIBERNIA's vales;[46]

I assumed that the *Anti-Jacobin* writers would pounce on Darwin's evident enthusiasm for the American and French Revolutions. Sure enough: on

looking up 'The Loves of the Triangles', I discovered that they had invented
a female parallel to his warrior of Liberty—'Young Freedom'. The anti-
French example they used was the massacre at Lodi two years earlier, when
Napoleon's army had slaughtered Austrian soldiers retreating through Italy:

> So, towering Alp! from thy majestic ridge
> Young Freedom gazed on Lodi's blood-stained *Bridge*;[47]

Returning to Darwin, I embarked on the next section, an extended passage of metaphorical imagery and biblical allusion. Just as subterranean
gnomes taught volcanoes to force open dark granite caves, and just as
Franklin released electrical charge trapped inside stormy clouds, so too,
Liberty sets free the prisoners of the Bastille and rips off the French nation's
repressive chains:

> —Touch'd by the patriot-flame, he rent amazed
> The flimsy bonds, and round and round him gazed;
> Starts up from earth, above the admiring throng
> Lifts his Colossal form, and towers along;
> High oe'r his foes his hundred arms He rears,
> Plowshares his swords, and pruning hooks his spears;[48]

As a parodic counterblast, 'The Loves of the Triangles' emphasized Napoleon's
imperial ambitions and the mass guillotining in post-Revolutionary Paris:

> Bade at her feet regenerate nations bow,
> And twined the wreath round Buonaparte's brow.
> —Quick with new lights, fresh hopes, and alter'd zeal,
> The slaves of Despots dropt the blunted steel.[49]

At this time, Napoleon was drawing up plans to cross the Channel and
invade England. To drive home their message, the satirists depicted the
worst-case scenario—the French navy would succeed, and destroy the
British way of life:

> Nor long the time ere Britain's shores shall greet
> The warrior-sage with gratulation sweet . . .

—The Communes spread, the gay Departments smile,
Fair Freedom's Plant o'ershades the laughing isle.[50]

As well as political revolution, Darwin also gives natural evolution his poetic treatment. Revolving around a central theme of change, this canto presents an overt challenge to the Christian concept of a stable universe created once and for all by God at the beginning of time. In Darwin's developmental model of the Earth, continents are heaved up above the primal ocean, marine shells give rise to limestone rocks, vegetables decompose into iron, lead transmutes into silver, and people band into societies that remove the forests once covering the surface of the globe.

To explain these transformations, Darwin proposes that matter is perpetually circulating. Following Lucretius, he explains that atoms are constantly brought together in different ways to form new minerals, animals, and vegetables. Although his theory is only vaguely formulated, it shocked his contemporaries by excluding any reference either to God's life-giving intervention at birth or to the soul's survival after death. Instead, according to Darwin, some ill-defined yet all-pervading spirit is recycled from one being to another. It is here that he introduces the controversial term 'Ens' (later to reappear in *Zoonomia*), by which he means an abstract entity or being that is somehow crucial for existence:

> GNOMES! with nice eye the slow solution watch,
> With fostering hand the parting atoms catch,
> Join in new forms, combine with life and sense,
> And guide and guard the transmigrating Ens.[51]

Seeking illumination, I started to read the long note attached to 'transmigrating Ens'. The first thirteen words resembled modern beliefs in the conservation of matter: 'The perpetual circulation of matter in the growth of vegetable and animal bodies...'. But the rest seemed extraordinary to me. First comes a discussion of Pythagoras' belief in the transmigration of spirits—what some religions call reincarnation; next a claim that the universe contains a small amount of matter surrounded by an all-pervasive spirit; and finally the story that 'Jupiter threw down a large handful of souls upon the earth, and left them to scramble for the few bodies which were to be had.'

Mystified, I read on.

Canto III: Water

Embarking on my third canto, I anticipated correctly that a mixed selection of subjects would be included—the Nile overflowing, medicinal springs, Iceland, Jupiter and Juno, sea horses, marshes, Hercules (again), fire engines, and so on, all linked together by aquatic nymphs and maids. As his prefatory Argument makes clear, Darwin uses fluid examples to reinforce his circulatory model of the universe: just as blood flows around the body, so too, he argues, there is a continuous cycle in watery nature as steam rises from the oceans to be condensed into rain that replenishes the rivers flowing down to the seas.

Neither the mechanical precision of the first canto nor the political outrage of the second feature in Darwin's 590 lines on water. Slightly shorter than the other three, it has substantially fewer notes at the back of the book, and no direct tributes to Lunar friends. After reading it through, my intuitive impression was that Darwin had taken time off from his self-imposed task of advertising his colleagues' achievements, and instead enjoyed luxuriating in composition. Although not to my own taste, the poetry seems more fluent and lyrical as it moves between imagined semi-mythical worlds and Darwin's immediate everyday experiences. For example, very near the end, he draws an extended comparison between children skating across a frozen pond and camel caravans traversing a desert, linking them through this familiar, well-observed image:

> So shoot the Spider-broods at breezy dawn
> Their glittering net-work o'er the autumnal lawn;
> From blade to blade connect with cordage fine
> The unbending grass, and live along the line;[52]

The most remarkable feature of this canto is the striking black-and-white illustration designed by Fuseli and engraved by Blake. Captioned 'The Fertilization of Egypt', it shows Anubis, the jackal-headed Egyptian god who protected the dead in the afterlife. Stretching his arms up towards a blazing light in the heavens, he straddles a river being replenished by a waterfall. In a long note, Darwin summarizes French reports that Sirius, the Dog Star, rises when the Nile's annual flood begins, and figures of Anubis are hung up in the temples. Peering through Anubis's wide-spread legs is a bearded face, an early representation of Urizen by Blake, often said to have been influenced by Darwin's receptive approach to mythology.

When Darwin writes about the ocean or foreign countries, his style tends to be vague and grandiose. It occurred to me that although Darwin spent much of his time travelling, he may never have seen the sea. He lived in the Midlands, and it was not until the Victorian age that the coast was regarded as a suitable place for holidays. With justification, many people suspected water to be a source of infection: when Franklin succumbed to a bout of fever after visiting Darwin, he blamed 'the Amount of Dabbling in and over your Ponds and Ditches ... after Sunset, and snuffing up too much of their Effluvia'.[53] Darwin did, however, enjoy river swimming, and there is presumably some foundation to the story that he once leapt off a boat after an alcohol-fuelled picnic, swam to shore, and then walked over the fields to Nottingham—where, unhindered by his usual stammer, he delivered a lecture on the benefits of fresh air.[54]

In contrast, at other points in this canto, Darwin's verses seem soundly based on his personal experiences. In this description of a pump, Darwin explains not only the mechanics of its operation but also why the water should foam:

> Bade with quick stroke the sliding piston bear
> The viewless columns of incumbent air;—
> Press'd by the incumbent air the floods below,
> Through opening valves in foaming torrents flow,
> Foot after foot with lessen'd impulse move,
> And rising seek the vacancy above.[55]

Buy why, I wondered, did he suddenly switch to breastfeeding for the next line? I found the answer in the note, which explains how babies learn to suck. There he adopted a clinical detachment very different from the tenderness with which this father of fourteen children by three partners portrays the little 'Plunderer' seeking 'the salubrious fount' of the fond mother's 'pearly orbs'. The vocabulary may be overblown, but the sentiment feels real.

Although lacking Lunar innovators, this canto is packed with local heroes: James Brindley, Darwin's diabetic patient whose canals had done so much to promote Midlands industry; an otherwise unidentified Miss Jones who secretly distributed her fortune to charitable causes; the Duke of Devonshire, who had developed his property at Buxton; and Mrs French, a fellow botanical enthusiast whose tomb beside the river Derwent was haunted by her grief-stricken husband. In only one passage does Darwin become carried away by an emotion matching the intensity of the previous canto: when describing a fire that had killed many children and a honeymoon couple, Darwin vehemently berates the water nymphs for failing to quell the flames and save their lives.

Darwin did slip in at least one incendiary passage. These apparently innocuous lines come immediately after a gushing tribute to the Duke of Devonshire and his ornamental fountains at Chatsworth House:

> NYMPHS! YOUR bright squadrons watch with chemic eyes
> The cold-elastic vapours, as they rise;
> With playful force arrest them as they pass,
> And to *pure* AIR betroth the *flaming* GAS.[56]

Well-informed readers would have realized that one of London's leading experimenters was Henry Cavendish, the eccentric scion of the Devonshire family who had isolated hydrogen, known as inflammable air. To loyal Britons, chemistry represented revolution, explosion, and French republicanism, so that merely to mention gases was to make a subversive statement. The focus of French research, gases were both literally and politically explosive—and to make it worse, Darwin's notes include an exceptionally early reference to oxygen, the chemical whose name had recently been invented in Paris. In the science of the 1780s and 1790s, burning was the burning question. Was the crucial ingredient dephlogisticated air, as maintained by the Lunar Society chemists Joseph Priestley and James Keir, or was it oxygen, in reality the same gas, but the theoretical brainchild of Priestley's Parisian rival, Antoine Lavoisier?

Looking back, phlogiston seems a ridiculous imaginary substance, but conceptually, it remained useful for around a century. Weightless and invisible, phlogiston was invented by German mining chemists to explain how smelting works. When metallic ores are heated with charcoal, they absorb phlogiston and turn into pure metal (in modern terms, oxygen is released from an oxide); conversely, when metals are heated (that is, they absorb oxygen), they release their phlogiston, which can sometimes be seen as a blue sheen on the surface. Problems started when chemists introduced more accurate balances, which showed that metals gain weight when heated—surely phlogiston could not have a negative weight? The answer seems more obvious now than it did at the time. Depending on which side of the Channel you lived, Lavoisier had either plagiarized Priestley's results or modernized chemistry by making it rational. For an English writer to endorse oxygen was to take a political stance by supporting French innovation over British tradition.

Darwin chose to write about gases in a canto focusing on water. Somewhat paradoxically, although he is presenting the latest theories, his poem is structured by the four Aristotelian elements, long discarded as a scientific explanation for what really exists in the world. Furthermore, Lavoisier had recently shown that water is not an element (of any kind), but a combination of oxygen and hydrogen. In his verse, Darwin deploys an old-fashioned vocabulary of marriage between '*pure* AIR' and '*flaming* GAS', but in the notes he uses Lavoisier's terminology.

Canto IV: Air

By now, I had a better idea of how to navigate round Darwin's poem. As I expected, in his fourth canto he blends together scientific innovators (notably his colleagues at the Lunar Society), mythical figures from the classics, and individuals who had recently hit the headlines for some particular feat or misfortune. Skimming through the summarized contents in the Argument, I learnt that I would yet again encounter an esoteric mixture of monsoons, Cupid riding a lion, balloons, Senacherib's army, crocodiles, insects, Kew Gardens, and Hygeia (think hygiene). I started to read.

Unlike the previous canto on water, Darwin here favours precise descriptions of household instruments and familiar situations. Whatever one may feel about their style, these instructions on grafting originate from an experienced gardener's pen, and (my poet informed me) are reminiscent of Virgil's *Georgics*:

> On each lop'd shoot a softer scion bind,
> Pith press'd to pith, and rind applied to rind,
> So shall the trunk with loftier crest ascend,
> And wide in air its happier arms extend;
> Nurse the new buds, admire the leaves unknown,
> And blushing bend with fruitage not its own.[57]

Darwin also once again treads on more controversial territory by celebrating air balloons, invented in France and linked with revolutionary ideals of freedom and equality. For the first time, French citizens had been able to soar above the king's head, and in 1810 the poet Percy Bysshe Shelley (then an unknown undergraduate) predicted that 'intrepid aeronauts' would survey Africa's vast interior from the air. 'The shadow of the first balloon,' he promised, 'as it glided silently over that hitherto unhappy country, would virtually emancipate every slave, and would annihilate slavery for ever.'[58]

In another nationalistic flourish, Darwin fantasized that Priestley's gases would enable Britain to rule the waves from below with 'Huge Sea-Balloons':

> The diving castles, roof'd with spheric glass,
> Rib'd with strong oak, and barr'd with bolts of brass,
> Buoy'd with pure air shall endless tracks pursue,
> And Priestley's hand the vital flood renew.—
> Then shall Britannia rule the wealthy realms,
> Which Ocean's wide insatiate wave o'erwhelms;[59]

Speaking through the Goddess of Botany, Darwin portrays Nature—not God—as the permanent regenerative force:

> With Life's first spark inspires the organic frame,
> And, as it wastes, renews the subtile flame.[60]

In Darwin's vision, the universe is characterized by conflict as well as by growth. As a doctor, he had witnessed many deathbed scenes, and in his letters he wrote eloquently and with great distress about his failure to save sick patients, especially children. Recalling the violent imagery of the second canto, Darwin describes the perpetual onslaught of contagious vapours, reeking marshes, and pestilential winds that mirror the horror of human battlefields:

> Hark! o'er the camp the venom'd tempest sings,
> Man falls on Man, on buckler buckler rings;
> Groan answers groan, to anguish anguish yields,
> And DEATH's loud accents shake the tented fields![61]

At this point, serendipity intervened, in the form of a book review about mock-epic poetry. One of its quotations sounded very familiar—and on checking, I realized that here Darwin is emulating some lines in Pope's translation of Homer's *Iliad*:

> Now Shield with Shield, with Helmet Helmet clos'd,
> To Armour Armour, Lance to Lance oppos'd.[62]

Eighty lines later, Darwin has reverted to a Lucretian style, describing how Nature will win and survive. Although the stars will fade like flowers in a field, they will be outlived by an eternal Nature emerging triumphant when all about her is chaos:

> —Till o'er the wreck, emerging from the storm,
> Immortal NATURE lifts her changeful form,
> Mounts from her funeral pyre on wings of flame,
> And soars and shines, another and the same.[63]

As a grand finale, Darwin closes this canto (and so the entire poem) by reiterating his own personal mission in life: to improve the health of the British nation. The last goddess he introduces is Hygeia, who is enticed by Botany's sylphs to a shrine decorated with flowers. These are the Goddess of Botany's parting lines:

> Oh, wave, Hygeia! o'er Britannia's throne
> Thy serpent-wand, and mark it for thine own...
> Shed o'er her peopled realms thy beamy smile,
> And with thy airy temple crown her isle![64]

And with that, she climbs into her silver carriage and whisks herself away. Her imaginary visit to Darwin's botanic garden is over.

I closed the book, and wondered what to make of it all. And what to do next.

I had one major poem left to read—the posthumously published *Temple of Nature*. I knew that contained Darwin's boldest speculations about the origins of life and the development of human society, and despite my lack of enthusiasm for heroic couplets, I was looking forward to reading it. But I wanted to be well prepared. In particular, I was intrigued by the thought that Darwin's belief in evolution was related to his campaign against slavery.

I discovered that most books about the abolition movement start in the early nineteenth century, which was after Darwin's death. If I wanted to think about the closing decades of the previous one, when he was thinking and writing, then I needed to investigate Enlightenment slavery for myself.

CHAPTER NINE

The Triangular Slave Trade

'Twas mercy brought me from my pagan land,
Taught my benighted soul to understand
That there's a God, that there's a Saviour too:
Once I redemption neither sought nor knew.
Some view our sable race with scornful eye:
'Their colour is a diabolic dye.'
Remember, Christians, negroes black as Cain
May be refined and join th'angelic train.

Phyllis Wheatley, *On Being Brought from*
Africa to America (1773)

When you order up an old book, you never know exactly what the librarian will bring you. Sometimes an item has disappeared, supposedly lost during a war or perhaps dispatched unbelievably long ago to the binding department for repairs. If it does turn up, you may discover that it is irrelevant or a reprint of something else, but with a new title. Sometimes sections have been bound in upside down or in the wrong order by a tired craftsman; more often, crucial pages have been excised by collectors or damaged by accidental spills of ink. Very, very occasionally, you stumble across some handwritten notes, jotted down centuries ago, that to anyone else would seem trivial—but as if destined to be found by you alone, immediately illuminate the quarry you have been pursuing. The intellectual thrill of sudden comprehension when half-triggered clues click into place is enormously gratifying.

One day, I unwittingly requested what turned out to be a uniquely informative copy of *The Botanic Garden*.[65] The beautiful leather binding stamped with a golden coat of arms suggested that this particular volume had originally been bought by an aristocrat not to read, but to show off to his friends—rather like all those unopened copies of Stephen Hawking's *A Brief History of Time*. Carefully tucked inside were several sheets of comments by one of Darwin's admirers, the poet William Cowper. Such chance discoveries still occur, and they can send research projects spinning in unanticipated directions. (Months later, at a seminar on furnishing fabrics, I met a Cowper scholar who gratefully dashed off to explore my serendipitous find.)

As I leafed through the yellowing pages, I realized that Cowper was making preparatory notes for a review. Unlike now, eighteenth-century literary reviews were generally anonymous and comprised collections of extracts strung together with a few comments. One scribbled remark in particular leapt out at me: 'The blood of the Mexicans shed by the Spaniards, and of the Africans by ourselves finaly [*sic*] resulted in the following indignant lines.' Seizing on a political rather than an evolutionary implication of Darwin's poem, Cowper was referring to these lines from the dark second canto of *The Economy of Vegetation*:

> Heavens! on my sight what sanguine colours blaze!
> Spain's deathless shame! the crimes of modern days!
> When Avarice, shrouded in Religion's robe,
> Sail'd to the West, and slaughter'd half the globe;
> While Superstition, stalking by his side,
> Mock'd the loud groans, and lap'd the bloody tide...
> Hear, oh, BRITANNIA! potent Queen of isles,
> On whom fair Art, and meek Religion smiles,
> Now AFRIC's coasts thy craftier sons invade
> With murder, rapine, theft,—and call it Trade![66]

By 'Avarice, shrouded in Religion's robe', Darwin meant Hernán Cortés, the despotic, gold-hungry leader who imposed Catholicism on Mexico in the 1520s after slaughtering thousands of Aztecs and smashing their religious shrines.

What particularly intrigued me (and Cowper) was Darwin's jump from Mexico to Africa. Cowper's own poem, 'The Negro's Complaint' (1788), had aroused immediate support for the anti-slavery movement four years earlier. Writing in the voice of a black African, Cowper had demanded to know how a Christian God could sanction the cruelty inflicted by English traders on fellow human beings haggled over for 'paltry gold'. From the abolitionists' perspective—Darwin points out here—British merchants were guilty of crimes as heinous as those of the Spanish conquistadors, disguising rapacious plunder as mutually beneficial trade.

With this accusation in mind, I reread some of Darwin's correspondence. When he sent *The Economy of Vegetation* to Wedgwood, he asked him to read only two sections: 'the page on the Slave-trade 117', which had been a last-minute addition, and his eulogy to John Howard (the prison reformer).[67] As a confirmed abolitionist, Cowper immediately picked out Darwin's anti-slavery sentiments. In contrast, although they had been there for me to see, I had overlooked them, my mind primed by modern critics to think about Darwin's industrial initiatives. Could there have been slaves in eighteenth-century Britain, the supposed land of enlightened liberty? Was Darwin's support of abolition linked to his evolutionary ideas? I decided to pursue these new lines of enquiry.

Histories of the British abolition movement focus on William Wilberforce, caricatured in Figure 3 (see p. 112) as the thin man to the left. Here Gillray is referring to Wilberforce's attempt in 1796 to end the slave trade, when he was defeated in Parliament by a margin of four votes, accused of colluding with Pitt to divert public sympathy away from the hardship suffered in Britain by poor white labourers. Gillray cynically portrays him smoking what I take to be joints (the modern catalogue calls them cheroots) with a fat cleric, the Bishop of Rochester. That name conveniently linked him to the Earl of Rochester, whose pornographic book lies on the table—but could that be the only connection? Once I had found out that in non-ecclesiastical life Wilberforce's fellow libertine was Samuel Horsley, I managed to work out the answer.

Most books on slavery do not even list Horsley in the index, and I had encountered him before only as the Fellow of the Royal Society responsible for suppressing Newton's obsession with alchemy. But the caricature made me appreciate that at the time, he was known as one of Wilberforce's influential allies. Strongly opposed to the slave trade, Horsley scoured travellers' accounts to demonstrate that Africans were not the savages traders made them out to be. Counter-intuitively, he was also a diehard Tory, satirized as a Turkish mufti for insisting that ordinary people should have nothing to do with laws except obey them. A High Church extremist, he waged theological warfare against Priestley, believed that Jacobins had taken over Methodism, and likened the French Revolution to the Beast of the Apocalypse. Presumably Wilberforce had mixed feelings about finding Horsley on the same side.[68]

Eventually, in 1807, Wilberforce succeeded in getting an anti-slavery bill through Parliament. Although he became Britain's major hero of the abolition movement, his success owed much to the previous thirty years of campaigning. Often forgotten about, those people and events formed the backdrop to Darwin's poetry and the *Anti-Jacobin*'s parody.

Black Africans in Enlightenment Britain

If this were a work of historical fiction, I would have no qualms about conjuring up scenes of cosy domesticity in Darwin's Lichfield home during the autumn of 1771. When not rattling round the countryside in his carriage visiting patients, the doctor—so my scenario would run—spends much of his time happily ensconced in his study, working on inventions, writing to friends, and entertaining visitors. In the rest of the house, servants prepare the meals and do the cleaning while his sister looks after his three sons, helped by their governess Mary Parker, who is pregnant with Darwin's first illegitimate daughter.

So far, all this is true. But in a novel, I would be free to speculate that on a dreary night of November—and I would specify 30 November, to demonstrate the assiduity of my research—Darwin dropped in to the local auction at the Baker's Arms. This imagined Darwin (and perhaps the real one too) knew from advance publicity that the lots included 'A NEGROE BOY from Africa, supposed to be about ten or eleven Years of Age'. According to his advertisers, this slave suffered neither from malnutrition, ill-health nor rebelliousness; they described him as 'well proportioned, speaks tolerable good English, of a mild Disposition, friendly, officious, sound, healthy, fond of Labour, and for Colour, an excellent fine Black'.[69] If Darwin did attend, perhaps he was indignant that a child the same age as his own should be sold to the highest bidder. He might even have put in an offer for this boy, emulating his local rival Samuel Johnson, who treated his own Jamaican servant excellently.

Buying African slaves had long been officially sanctioned. In 1660, the year Charles II was restored to the throne after the Civil War, two new gentlemen's clubs were founded. One of them still survives—the Royal Society of London, now dedicated to science. Its founding Fellows boasted about their 'Twin-Sister', the Company of Royal Adventurers into Africa, which was given a monopoly over the British slave trade. 'In both these *Institutions* begun together,' explained the Royal Society in a flattering comparison of Charles and Solomon, 'our *King* has initiated the two most famous *Works* of the wisest of antient *Kings*: who at the same time sent to *Ophir* for *Gold*, and compos'd a *Natural History*, from the Cedar to the Shrub'.[70] In other words, commercial trade and scientific research would work together to ensure that Britain dominated the globe. As they predicted, by the end of the seventeenth century, the slave trade had made London the world's wealthiest city.

Britain was unusual not because it participated in the slave trade, but because it did not enslave its own citizens. Britannia ruled the waves, Britons never, never, never would be slaves—but this patriotic hymn of 1740 said nothing about owning slaves from other countries, said nothing about the armed press gangs who roamed the country kidnapping sailors for forced service, and said nothing about the thousands of British captives abroad. Losing personal liberty had little to do with being black: on one modern estimate, around nineteen out of every twenty people around the globe were enslaved. One lawyer argued against legalizing domestic slavery in Britain by warning that victims would 'come not only from our colonies and those of other European nations, but from Poland, Russia, Spain and Turkey, from the coast of Barbary, from the western and eastern coasts of Africa, from every part of the world, where it still continues to torment and dishonour the human species'.[71]

Industrialization made Britain rich, but marketing people was as important as inventing machines. When the 'Loves of the Triangles' appeared, the triangular slave trade linked Britain, Africa, and America in a continuous profit-generating circuit. Mostly sailing in the same clockwise direction as the prevailing winds and currents, ships travelled with packed holds, any human cargo stashed into cramped wooden frames resembling wine racks. On the African coast, British goods were exchanged not only for gold, cotton, and palm oil, but also for captives seized by local traders: the going rate was about one gun per person. On arriving in the Caribbean or Latin America, the imprisoned Africans (at least, those who survived) were unloaded onto the quayside and displayed for sale so that they could cultivate the plantations. Merely by arriving, these men, women, and children had effectively earned interest for their shippers—and as far as their new owners were concerned, it was cheaper to import replacements than to feed them properly. For the return trip across the Atlantic—to British ports such as Bristol, Liverpool, and London—ships were stocked with crops produced by earlier consignments of slaves. Sugar was by far and away the most important, but wealthy Britons wanted tobacco and also rice, the new luxury originally taken over to the Americas by enslaved Africans.[72]

This human trade gratified the British appetite for a sweeter diet and drove the processes of industrialization that Darwin hymned so eloquently in his poetry. Slavery was, wrote an economist urging the government to invest in Africa, 'the first principle and foundation of all the rest, the mainspring of the machine which sets every wheel in motion'.[73] During the eighteenth century, three million urban jobs were created in Britain, and rapidly expanding cities such as Manchester and Birmingham depended for their wealth on goods that could be bartered in exchange for people. British banking and insurance businesses boomed, Yorkshire textiles relied on African and New World markets, and the fine metalware produced in Birmingham's factories embellished wealthy American homes. Between 1783 and 1793, Liverpool's ships transported over three hundred thousand black slaves: citizens made rich showed their gratitude by decorating the Town Hall with carved heads of Africans and elephants.[74]

By 1760, the black population of London was large—probably around fifteen thousand, although accuracy is impossible. Best-off and best-fed were those arriving as the enslaved servants of visiting plantation owners who were economizing by not hiring expensive local help.[75] Many were newly imported or second-generation slaves, often badly treated, or else escapees condemned to life on the run: hunting down and selling escaped slaves was legal as well as profitable. Desperately poor, these outcasts—mostly male—banded together for protection in tightly knit communities.

Often branded with patronizing classical names—Pompey, Ovid, Socrates—enslaved Africans were identifiable not only by their colour but by their collar. In 1756 a London goldsmith advertised 'silver padlocks for Blacks or Dog; collars &c.'[76] In other words, owners could legitimately treat their black possessions as though they were animals. By chance, when gazing at a museum case of porcelain shepherdesses and cows, I once spotted Aesop, an African who conformed to ceramic convention by sitting on a tree stump strewn with flowers. Although wearing a white collar and bright pink jacket, he carried a chain around his neck, and his hunched shoulders were designed to make him look bestial.[77]

Like Samuel Johnson, some families cared generously for their black servants. However, many indigenous Britons resented these immigrants, justifying their prejudice with depressingly familiar arguments. Top of the list were economic excuses. After all, ran the rhetoric, when there was already too much competition from French and Irish settlers, blacks would take away still more jobs. Their children would be a burden to the state; in any case, blacks were—everybody knew this, of course—physically and temperamentally better suited to hot climates. They were also less diligent, healthy, and intelligent than white workers, to say nothing of being intrinsically dishonest. And if they managed to seduce a fair-skinned bride, they would pollute/stain/contaminate the island race.[78]

The British Abolition Movement

By some quirk of fate, the 1771 newspaper advertisement for the Lichfield auction that Darwin may or may not have attended happens to have survived. Presumably, other small boys were being sold in towns all over the country. Yet only a couple of years later, marketing slaves had become illegal in Britain, and protesters were campaigning for abolition. What happened, and why then, I wondered? After such a long period of acceptance, why should slavery suddenly have become unacceptable?

I soon stopped believing in those wishful-thinking narratives that portray abolition as inevitable progress towards civilized equality. According to them, Enlightened liberty was sweeping across Europe, whereas in reality most people continued to put their own interests first. Slavery underpinned the British economy, and banning it would have had enormous financial implications. To many, the existing order of things seemed indubitably right. For once turning against his idolized Johnson, James Boswell protested that 'To abolish a *status*, which in all ages GOD has sanctioned, would be *robbery* to an innumerable class of our fellow-subjects.'[79] Boswell spoke for many.

Of course, attitudes were gradually changing under the impact of new ideas. The educational theories introduced by John Locke at the end of the seventeenth century suggested that black Africans could—like women, labourers, and children—acquire the skills previously reserved for white males. Explorers published romanticized accounts of their experiences overseas, making it clear that the British system of government was not the only one that worked. Instead of characterizing people by their religion or clothes, Europeans were starting to place greater emphasis on physical characteristics such as skin colour and physical strength. Debates about morality and civic responsibility prompted more sympathetic treatment of less-privileged groups—slaves and servants, children and animals, non-Anglicans, and prostitutes. But appreciating these long-term shifts does not explain why there were so few debates and articles about slavery until 1788. Why did they increase so suddenly?

The more I read, the more I became convinced that serendipity rather than ideology precipitated the anti-slavery movement in Britain, and that religious commitment was just as significant as political radicalism. A small kernel of abolitionists rapidly influenced the entire country by taking advantage of Britain's modern transport system and uncensored press. The movement was sparked by evangelical Christians who experienced sudden conversions to this new cause, and sustained by hard-working Quakers, already committed to equality, who used their pre-existing business skills to organize a nationwide movement. Before Wilberforce became involved, two other pioneers, now less famous, were crucial—Granville Sharp and Thomas Clarkson. Their biographies contain as many coincidences, chance encounters, and unanticipated events as a weak detective novel.[80]

Sharp's nearest modern equivalent would be an immigration lawyer acting on behalf of asylum seekers, but he had never envisaged such a career for himself. In his previous existence as a junior civil-service clerk, Sharp spent much of his time on a barge playing music with his extended family. This was no ordinary barge—its orchestral concerts were so famous that even George III and his wife came on board to listen. One day in 1765, Sharp's life jumped tracks. Calling in at his brother's house, Sharp found him looking after an African teenager called Jonathan Strong, who had been severely injured by his owner and then left on the street. The Sharp brothers restored Strong to health, found him a job as a footman, and perhaps thought little more about their generosity.

Two years later, another coincidence intervened. Strong's former master bumped into him accidentally, seized what he regarded as a valuable item of his own property, and sold him to a Jamaican planter. Sharp managed to rescue Strong—probably helped by having once entertained the Lord Mayor on his musical barge—and from then on, financially supported by his family, set out to make himself Britain's leading expert on slavery. He soon discovered that the laws were vague. In principle, everybody in Britain had a natural right to freedom, but there was no ban on trading slaves internationally. Moreover, many indentured servants were effectively captives, unsalaried and with no means of surviving away from the home where they found themselves working.[81]

Sharp's big opportunity came when he took on the case of James Somerset, a black slave who had escaped from his Boston master during a trip to London but was seized and clapped in irons so that he could be sent back to the plantations. Accusing his owner of assault, some public-spirited witnesses engineered Somerset's release, but after months of delay, the judge wriggled out of making any definitive pronouncement. In 1772, in a fence-sitting decision that effectively allowed slave-trading to continue until 1834, he issued only the half-hearted, ambiguous decree that escaped slaves could not be forcibly recaptured for sale and sent abroad. To own slaves in Britain was illegal—but nothing prevented British people from buying or transporting them.

This paradoxical judgement was ridiculed by the lawyer Thomas Day, Darwin's friend at the Lunar Society. 'I cannot say I perfectly understand', he wrote sarcastically in an anti-slavery tract, 'how the negroe is a free man, and yet his master may possibly have a right to his service.'[82] An unusually early literary protestor, the following year Day published his poignant poem (*The Dying Negro*) about a black slave bidding farewell to his fiancée, suicide gun in his hand. He promoted still further public out-cry with *A Letter on the Slavery of the Negroes* (1785), which describes the notorious affair of the *Zong*.

In 1781 the Liverpool-based *Zong* lost its way in the Caribbean, and as water supplies ran low, the captain ordered 133 sick slaves to be thrown overboard. The names of the drowned remain unknown: in a trial focusing on money not morality, the victims were recorded merely as numbers and financial values. In his extraordinary defence, the captain argued that he had been acting in his employer's best interests. If the slaves had died natu-rally, he maintained, the loss would be borne by the shipowner, whereas because he had acted promptly in an emergency to save the rest of the pas-sengers, the insurers were liable. At the trial a couple of years later, Sharp objected strongly but had little impact on the chief judge, who denied that murder was involved because offloading African captives was the legal equivalent of killing horses.

Like Sharp, Clarkson wandered into political activism almost uninten-tionally. Endowed with striking red hair and over six feet tall (a more impressive height then than now), this Cambridge graduate spent most of his life touring Britain to deliver dramatic lectures on the evils of slavery. Reading between the lines of his dry biographies, it seems that as a student he had spent much of his time partying, gambling, and riding his two horses. Even so, after graduating in mathematics he decided to stay on at Cambridge and become a clergyman, but started entering Latin essay competitions—less bizarre a pastime than it might seem, since young gen-tlemen were expected to be fluent.

Asked (in Latin) 'Is it lawful to make slaves of others against their will?', Clarkson won first prize. The consequences were unexpected. A small monument in Hertfordshire still marks the spot where Clarkson paused to rest during a horse-ride to London and—as Coleridge put it—became a 'moral Steam-Engine'.[83] In a sudden conversion by the roadside, Clarkson decided that halting the Atlantic slave trade mattered more to him than anything else. Embarking on a one-man, single-minded mission, he published his prize-winning essay, translated into English to reach a wide readership. Rallying supporters, he managed to recruit the MP William Wilberforce—a far more significant step than he realized at the time—as well as enlisting the financial backing of wealthy industrialists. In particular, Darwin's Lunar colleague, the self-made factory-owner Josiah Wedgwood, joined the organizing committee of Clarkson and Wilberforce's Society for the Abolition of the Slave Trade.

As Clarkson's proficiency in Latin suggests, orators and lawyers traditionally relied on learned texts, legal precedents, and rhetorical speeches. In addition to lobbying potential allies and publishing articles, Clarkson introduced what has now become a common procedure of persuading through showing. Like a Baconian experimenter, he believed in the power of objects to shock and convince—and like an itinerant scientific lecturer, he travelled round the country displaying objects that silently reinforced his words. Although he never visited Africa himself, over the years Clarkson accumulated many examples of grains, spices, and handicrafts to provide immediate, visible evidence that black slaves were not intrinsically inferior, but had been wrenched out of a rich culture and a fertile continent. To gather further information, he boarded large trading ships and mingled with immigrants in the ports, often risking his own life as he tracked down evidence of deaths that captains were trying to conceal.

Pursuing Clarkson, I visited his birthplace—Wisbech, a small town in the Fens. He knew it as a thriving inland port with elegant eighteenth-century mansions, but those that survive are now overwhelmed by cheap take-aways and tattoo parlours. It was the town's tiny museum that surprised me most: its faded display cases contain not only mangled coins and fragments of Roman pottery, but also medical equipment (the bedpan collection is particularly splendid) and African artefacts (including some extraordinarily skimpy beaded skirts). And it was here that I saw Clarkson's 'African Box', familiar to me from his own book, and also from a fine portrait on display at Wilberforce's house in Hull.

Resembling the specimen chest of a botanical or mineralogical collector, this Box accompanied Clarkson on his tours around Britain, acting as a vital stage prop in his dramatic talks. It was fitted with a detachable wooden tray subdivided into small lidded compartments containing samples of African foodstuffs—pepper, millet, tamarind, black-eye peas. To demonstrate that Africans were as capable as Europeans of being artistic and inventive, hard-working and intelligent, Clarkson also piled up pieces of craft work—woven cloths, decorative baskets, leather necklaces.[84]

During his dockside investigations, Clarkson collected some very different items to impress his listeners—shackles, cat-o'-nine-tails, handcuffs, all made for profit in Britain's flourishing Midlands metal factories. When placed next to the skillfully crafted African cloths and statues, they must have looked embarrassingly barbaric. Darwin endorsed Clarkson's strategy of display. 'I have just heard', he wrote to Wedgwood in 1789, 'that there are muzzles or gags made at Birmingham for the slaves in our islands.—if this be true, & such an instrument could be exhibited by a speaker in the house of commons, it might have great effect.—could not one of their long whips, or wire-tails be also procured, & exhibited? But an instrument of torture of our own manufactory would have a great effect I dare say.'[85]

If Darwin ever attended one of Clarkson's lectures, he might have seen these devices (Fig. 5), products of Britain's ironworking industry that Clarkson had bought after spotting them in a Liverpool shop window. The vendor explained in gruesome detail how to inflict the maximum pain and discomfort. At the top is a pair of handcuffs designed to padlock a slave's wrists close together; the leg clamps at the bottom were designed to fasten one slave's right ankle to the left foot of his or her neighbour (although some naval tyrants increased the agony by crossing their legs over first). As Clarkson observed, the existence of such restraints demonstrated that when Africans 'were in the holds of the slave-vessels, they were not in the Elysium which had been represented'.[86]

The functions of the other two objects are less immediately obvious, and if reproduced in a book about the history of medicine, might well be invoked to illustrate the sophistication and precision of Britain's instrument trade in the late eighteenth century. They ensured that, should standard shackles and whips prove insufficient as deterrents, captains could still quash rebellious protests. The device with two hoops is a thumb-screw, while the one at the centre right was familiar to surgeons as a *speculum oris*. Originally designed for treating lockjaw, it was ideal for

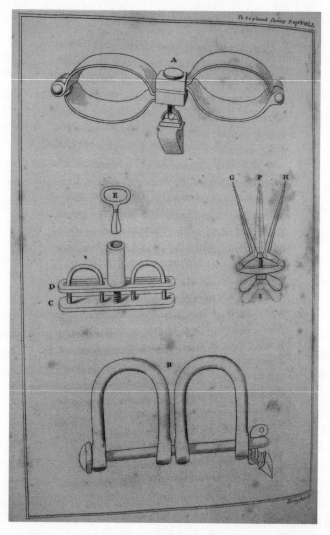

Figure 5 Slavery metalware collected and illustrated by Thomas Clarkson. Plate from Thomas Clarkson (1808), *The History of the Rise, Progress, and Accomplishment of the Abolition of the African Slave-Trade by the British Parliament.*

force-feeding slaves who preferred to die through starvation than continue across the Atlantic towards captivity, hard work, and an early death.

By converting instruments of control into empirical pieces of evidence, Clarkson made slaves visible in England. Instead of being concealed as abstract entries in the credit columns of a ledger, they were now on display as human victims.

The Economy of Abolition

The abolitionists wrung hearts by exposing cruelty, but they achieved change by applying economic pressure. In the 1790s Britons were consuming 4 kilograms of sugar per person, the highest annual rate in Europe, which not only brought profit for plantation owners but also generated tax income for the government. In an early equivalent of the Free Trade movement, campaigners urged consumers to show their support for the anti-slavery movement by refusing to buy imports from the West Indies.

Propagandists called tea 'the blood-sweetened beverage'—and this was sometimes literally true. The cone-shaped sugar blocks that crossed the Atlantic contained blood, sweat, and even flesh, because workers sustained deep cuts from the sharp cane leaves, and arms or legs got trapped in the processing machinery. To eat white sugar, argued the abolitionists, was the inverse of black cannibalism, and to intone a Christian grace over a blood-soaked meal was an act of sacrilege.[87] Darwin recommended growing British sugar beet as a morally clean source of much-needed calories: 'the slaves in Jamaica grow fat in the sugar-harvest, though they endure at that time much more labour,' he wrote in his book on agriculture, before expostulating 'Great God of Justice! grant that it may soon be cultivated only by the hands of freedom.'[88]

The nationwide protest was so effective that sugar sales dropped by a third, tax revenues were slashed, and the government was forced to take notice. This was an unprecedented example of public action, taken not so much by the privileged as by the disadvantaged: this was a unique opportunity for women, black immigrants, and labourers to express their sympathy for fellow sufferers—and for their voices to have an impact. Never before had ordinary people made their opinion count so effectively, and never before had women who controlled the household purchases taken such decisive action. Husbands might complain about their unsweetened tea, but their wives and children remained resolute.

Clarkson and his colleagues solicited support by distributing emotive images, but they were designed for eighteenth-century sensibilities. Modern charities operate by promoting empathy with individual Africans and their fate, but two hundred years ago the strategy was to denounce white cruelty. Enslaved Africans were stripped of their identity, labelled with numbers by their purchasers and with new names by their owners. The route to British hearts lay in presenting these alien beings not as equals but as symbols of white sin. Britons were more concerned with their own moral salvation than with the fate of others: whether it concerned Africans, animals or women, those who treated inferior beings humanely could feel confident about their personal status as civilized Europeans.[89]

The abolitionists' publicity material concentrated on the 'middle passage', the mass transport of shackled slaves in cramped ships from their African homes westwards across the Atlantic. When on board British ships, the captives were British responsibility, and so, whatever their subsequent fate, should be cared for properly. Two images proved particularly potent. One was a diagrammatic cross-section of a ship, drawn with mathematical precision, depicting lookalike creatures packed in with maximum efficiency as though they were pieces of machinery, or animals in Noah's ark (Fig. 6). The large section nearer the bow is reserved for men, chained together by their feet. The boys are in the middle, separated by the women from their sisters crammed in to the stern between the store rooms. The other icon of abolition was Britain's first promotional logo—a kneeling handcuffed slave, encircled by the plea: 'Am I not a man and a brother?' (Fig. 7). Originally manufactured as a medallion in Wedgwood's distinctive pottery, the picture and the words rapidly became identified with the anti-slavery cause.

'Am I not a man and a brother?' This evocative slogan bears the ring of eternal truth, but its significance is different now from then, when brotherhood did not imply equality. At a literal level, brothers within the same family were unequal. This is still partially true in modern Britain, where

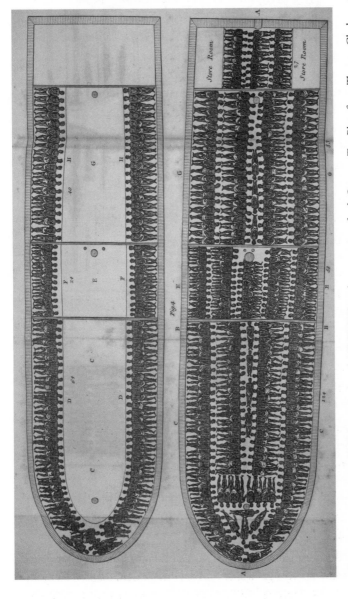

Figure 6 Plan of an *AFRICAN SHIP'S lower Deck with NEGROES in the proportion of only One to a Ton.* Plate from Thomas Clarkson (1808), *The History of the Rise, Progress, and Accomplishment of the Abolition of the African Slave-Trade by the British Parliament.*

Figure 7 'Am I Not a Man and a Brother', abolitionist medal, 1790s (copper), English School, (18th century). © Private Collection / © Michael Graham-Stewart / The Bridgeman Art Library.

the monarch's first son inherits the throne, and landowners keep estates intact by leaving them to the oldest boy. For his scene of an eighteenth-century London drawing room, the artist Arthur Devis divided his subjects into two groups. On one side are the father and one son—the one who stands to gain everything. On the other are the mother, the daughters—and a boy who looks small, but is, in fact, only minutes younger than his twin brother on the paternal side of the canvas. Playing with his sisters, this disinherited son is building a fragile house of cards, symbolizing that the luck of the draw can make or destroy a life.

Inequality was intrinsic to Britain's hierarchical society. Everyone knew their place, and just like a father looks after his children, the privileged were responsible for looking after those born into lower stations. The right to vote now seems one of the most fundamental in a democracy, yet all sorts of age and property restrictions were then in place. Only about one in twenty British men could vote—and, of course, not a single woman, however aristocratic she might be. Even in France, presented by Darwin as the land of change where equality reigned, brotherhood excluded sister-hood. *Liberté Egalité Fraternité* made an effective rallying-call for Revolu-tionaries, but it did little to alter the status of women. As Day pointed out, achieving equality in America also seemed remote. 'If there be an object truly ridiculous in Nature,' wrote Day, 'it is an American patriot signing resolutions of independency with one hand, and with the other brandish-ing a whip over his affrighted slaves.'[90]

In a late addition to *The Economy of Vegetation*, Darwin included an engraving of Wedgwood's captive African, remarking approvingly that his friend had 'distributed many hundreds to...assist in the abolition of the detestable traffic in human creatures':

> Form the poor fetter'd SLAVE on bended knee
> From Britain's sons imploring to be free;[91]

About a hundred lines later, he refers to it again:

> —The SLAVE, in chains, on supplicating knee,
> Spreads his wide arms, and lifts his eyes to Thee;
> With hunger pale, with wounds and toil oppress'd,
> 'ARE WE NOT BRETHREN?' sorrow choaks the rest;—[92]

Darwin was not simply repeating himself. Whereas the earlier couplet had contributed to his advertisement for Wedgwood's industrial success, this second description featured in his support for the French and Amer-ican Revolutions. As the caricature of the composite African/Jacobin (see Fig. 4, p. 130) indicates, abolition and political radicalism were seen as related issues. Many readers of the *Anti-Jacobin* agreed with the Earl of Abingdon, who explained in 1794 why he would vote against any anti-slavery legislation: 'this proposition for the abolition of the slave trade, is...grounded in and founded upon French principles...namely those of insubordination, anarchy, confusion, murder, havock, devasta-tion and ruin'.[93]

To reinforce the links between political radicalism and the abolition movement, 'The Loves of the Triangles' transfers Darwin's imagery of an elfin workforce from Wedgwood's factory to a French shipyard. These lines are from Darwin's flight of fancy:

> GNOMES! as you now dissect with hammers fine
> The granite-rock, the nodul'd flint calcine;
> Grind with strong arms, the circling chertz betwixt,
> Your pure Ka-o-lins and Pe-tun-tses mixt.[94]

The *Anti-Jacobin* parody is so finely tuned that it could almost have been written by Darwin himself. It's easy to imagine the young writers standing around the office table, perhaps slightly drunk, bent over with laughter as they goad each other into witty refinements. Here a cherub crew is building a boat for Napoleon to cross the Channel and invade England:

> Turn the stiff screw, apply the strengthening clamp,
> Drive the long bolt, or fix the stubborn cramp,
> Lash the reluctant beam, the cable splice,
> Join the firm dove-tail with adjustment nice,
> Through yawning fissures urge the willing wedge,
> Or give the smoothing adze a sharper edge.[95]

These fears of a French landing were very real at the time, but within a few years it had become clear that Napoleon would not invade Britain, and anti-Jacobin propaganda died down. In contrast, legal slavery continued for decades. When Darwin's grandson Charles arrived on the Brazilian coast in 1832, he was appalled to see Africans up for sale. 'The extent to which the trade is carried on,' he fumed in his diary; 'the ferocity with which it is defended; the respectable (!) people who are concerned in it are far from being exaggerated at home.'[96]

The abolitionists managed to prevent escaped slaves from being recaptured in Britain, but the economic benefits of international slavery outweighed the moral arguments against it. Banning the slave trade would have ruined not only Britain's traditional elite but also the newly rich factory owners of the developing Midlands. Darwin, Wedgwood, and other members of the Lunar Society were certainly not themselves implicated in making or selling the iron instruments that Clarkson displayed, yet their own affluence depended on industrialists who did. However genuine their hostility to slavery, they were also fallible, inconsistent human beings who struggled to balance their ideals, their responsibilities, and their pleasures.

Although ideologically committed to brotherhood, armchair philosophers campaigned from comfortable positions. Like most well-meaning people, they behaved inconsistently, a dilemma that still vexes me and countless other consumers trying to buy ethically. Enlightenment reformers were unwilling to sacrifice their own high standard of living, even though it relied on the exploitation not only of overseas slaves but also of British women and labourers who were underpaid and overworked. When Darwin celebrated Arkwright's looms in verse, he must have known that they were weaving cotton material for buying and clothing slaves. And when he reproduced Wedgwood's chained slave, he omitted to mention that this self-styled 'Vase-maker to the universe' had turned it to his own advantage by converting a political motif into a commercial product, a fashion item reproduced on snuffboxes and jewellery.

Other members of the Lunar Society also found themselves making compromises. Boulton and Watt seem to have been the most profit-hungry, and their reactions to the American Revolution wavered depending on the financial implications for sales. Priestley—dissenting preacher as well as industrial chemist and radical politician—took a pragmatic approach to abolition. When urging Christians to consider Africans 'as *brothers*, and *equals*', he added the consoling thought that, like all servants, they could simultaneously be regarded 'as *inferiors*'. If we were to liberate slaves now, Priestley explained reassuringly, they would be unable to cope with their unfamiliar freedom; far better that masters keep their current 'stock' and treat them kindly to get the most work out of them.[97]

Another Lunar man found himself in a particularly awkward position. Samuel Galton is rarely mentioned, but his young daughter later wrote some evocative memoirs of Darwin and the Lunar meetings. Although he was a Quaker espousing pacifism, Galton's family wealth came from manufacturing guns, that vital unit of currency in bartering for slaves.[98] When the Birmingham Friends debated the ethics of accepting a donation from Galton, he defended himself by pointing out that anybody who consumed food from the West Indies was guilty of indirectly supporting the slave trade: '...*those who use the produce of the labor of Slaves, as Tobacco, Rum, Sugar, Rice, Indigo, and Cotton [promote] the slave trade...because [their] consumption is the very Ground and Cause of Slavery...*'[99] (The Quakers disowned him, but they seem to have happily accepted his money after he abandoned his gun business and turned to banking instead.)

Whatever one might think of Galton's behaviour, he was surely right to emphasize the collective guilt of a nation whose entire economy depended on slavery. Explaining that silence is the easiest policy, William Cowper made Galton's point more eloquently, emphasizing his irony by choosing a metre (anapaestic tetrameter, my poet informed me) more often used for humour:

> I pity them greatly, but I must be mum,
> For how could we do without sugar and rum?...
> He blam'd and protested, but join'd in the plan;
> He shar'd in the plunder, but pitied the man.[100]

Darwin also acknowledged this clash between compassion and self-interest. In another poetic denunciation of slavery, he wrote that 'Inexorable CONSCIENCE...speaks in thunder..."HE, WHO ALLOWS OPPRESSION, SHARES THE CRIME."'[101] Darwin may have spent dark hours of the night examining the inconsistencies in his own position. But then again, perhaps he did not.

THE TEMPLE OF
NATURE

PROGRESS, RACE, AND
EVOLUTION

Technology...the knack of so arranging the world that we need not experience it.

Max Frisch, *Homo Faber* (1957)

I tried to imagine myself into the elegant buckled shoes of an eighteenth-century gentleman who liked to drop in at his local coffee house every week and peruse the *Anti-Jacobin*. Perhaps I would make a point of getting there early every Monday morning when the paper came out, to make sure that it had not mysteriously disappeared. And as I settled down in an armchair with my fashionable coffee or hot chocolate, I would probably turn first to my favourite section—the satirical poem. On 16 April 1798, six months after the journal was launched, I would have found myself contemplating the long prose introduction to 'The Loves of the Triangles'.

Pretending to reproduce a manuscript created by somebody else is a common literary device: famous examples include Jonathan Swift's *Gulliver's Travels* and Henry James's *The Turn of the Screw*. William Godwin—the imaginary author of 'The Loves of the Triangles'—was most famous for *An Enquiry Concerning Political Justice* (1793), in which he championed progress based on reason. Borrowing a word invented by the French philosopher Nicolas de Condorcet, Godwin introduced the notion of perfectibility. Like Darwin, he took it as given that human beings are continuously improving, so that progress is built into society. Unfortunately, he opened himself up to easy satire by over-optimistically predicting liberation from the shackles of physical mortality—a suggestion lampooned in the *Anti-Jacobin* as the '*eternal and absolute Perfectibility of Man*'.

In Godwin's opinion, whatever situation you find yourself in, you should always act to maximize the public benefit. So if a house is burning down, you should abandon the chambermaid in favour of rescuing her employer, the wise educator who will do more to improve the general good. And you should always tell the truth, even if it means offending your friends. Relentlessly pursuing this logical train of thought, Godwin argued that monarchs are unsatisfactory rulers because they are surrounded by sycophantic courtiers. Instead of making decisions, the king 'is paid an immense revenue only to dance and to eat, to wear a scarlet robe and a crown...he must have no opinion, but be the vacant and colourless mirror by which [his ministers' judgement] is reflected'. Since everything is constantly improving, he argued, the American and French Revolutions were better than previous uprisings—and Britain's next Revolution would be the best one yet.[1] Hardly surprising that the *Anti-Jacobin* was hostile.

Five years later, Godwin's frank biography of his wife, Mary Wollstonecraft, made him once again a controversial figure. His *Memoirs* of her appeared in January 1798; in April he was ridiculed in the *Anti-Jacobin*, and in July the first issue of the government-subsidized *Anti-Jacobin Review* carried a vicious critique of both Wollstonecraft and Godwin. In 'The Loves of the Triangles', Mr Higgins (aka Godwin) is presented as also being responsible for an earlier *Anti-Jacobin* three-part poem, 'The Progress of Man', whose first instalment had appeared in February. The ostensible target was Richard Payne Knight's *The Progress of Civil Society*, but Darwin and Godwin were also referred to by name.

These self-styled anti-Jacobins clearly found progress threatening. Nowadays it is so manifestly A Good Thing that the verb has become transitive: the contractors progressed the building, the manager will progress the company (forwards, even). But two hundred years ago, many people were convinced that change can take place in only one direction: for the worse. This gloomy prognosis was a central message of the Old Testament. When Adam yielded to temptation in the Garden of Eden, humanity lapsed from its original state of perfection, a symbolism that pervaded European culture—including science. Robert Hooke, for example, emphasized that he had invented his high-powered microscope to restore the defective vision of fallen mortals: only with that artificial aid could human beings perceive the beauty of God's creation at the minutest level.

At the end of the seventeenth century, new ideas still faced huge resistance. That was when the philosopher John Locke wrote: 'the imputation of Novelty, is a terrible charge amongst those, who judge of Men's Heads, as they do of their Perukes [Wigs], by the Fashion; and can allow none to be right, but the received Doctrine.'[2] Today's critics point to global warming, nuclear devastation, and chemical pollution as obvious downsides of advance, but this is a fundamentally different type of objection. Locke was referring to contemporaries who gazed back wistfully towards an ideal past, declaring that decline was unavoidable, almost as if preordained by God.

Progress has become a leitmotif of modernity, so ingrained that it seems instinctive.[3] Eighteenth-century writers inherited two approaches. One tradition stemmed from the Judaeo-Christian belief that God formed the world in a specific act of creation. In this version, time flies like an arrow, shooting out from its origin to leave the past behind as it travels towards the future. When twentieth-century cosmologists suggested that the universe had started with a Big Bang, this biblical story helped to make their model acceptable. There is, however, a crucial difference: whereas the Bible states that the universe was created as it is now, Big Bang theorists hold that it has been developing ever since its beginning.

On the other hand, many Greek philosophers envisaged a cyclical universe that keeps recurring, a model that also influenced Enlightenment writers. An Aristotelian cosmos exists in its own right, with no need for any external Divine Creator. A rosebud becomes a flower not because God ordained it, and not because it has consciousness, but because that is its purpose. Instead of directionality, there are no definite beginnings or ends but a never-ending sequence, rather like the annual seasons on a grand cosmic scale. This type of view also remained important. For instance, from studying Ovid's *Metamorphoses* (compulsory reading for any educated gentleman), naturalists were predisposed to regard shells embedded in the ground as evidence that lands and seas had changed places in the past.

Enlightenment authors were influenced by both these perspectives. To illustrate how civilizations can grow to a natural peak and then inevitably decline as a new one emerges, Edward Gibbon traced out the fall of the Roman empire under the impact of Christianity. But he also argued that progress was guaranteed for the foreseeable future, because printing ensured that knowledge could never be lost. Like so many seminal books, Gibbon's was successful not because it jolted readers out of their habitual patterns of thought, but because it articulated so eloquently the convictions they already held.

Few people nowadays would maintain that civilization has gradually declined ever since the Greeks, but that used to be a common point of view. Writers react to their own predecessors rather than anticipating successors. After Adam Smith laid down the principles of capitalist economics in the *Wealth of Nations* (1776), a fellow Scot scoffed that he 'has produced a book upon trade, from which one would think he had never read any of the writers of Greece and Rome'.[4] If you visualize humanity travelling through time as if through an undulating landscape, then on every downward slope it makes sense to look back for lost wisdom rather than assume inevitable progress towards a better future. When Samuel Johnson started work on his *Dictionary*, he set out to prevent further deterioration in the English language by fossilizing grammar and vocabulary in their current state (although he later concluded that this was impossible).

In the Victorian era, cyclical views about time still affected theories about the age of the universe. The most influential author on that topic was the geologist Charles Lyell, and he wrote for an audience already convinced by Enlightenment naturalists that the world must be far, far older than allowed for in the Bible. Lyell proposed that the world is constantly changing at the same slow, uniform rate, a concept he symbolized by describing an old Italian temple whose pillars had become striped with traces of marine life as they repeatedly sunk below the surface of the water and rose up again. Lyell's vision of gradual transformation effectively expanded pre-history, making it long enough to accommodate the time-consuming processes of evolution. By accepting Lyell's gradualism but ignoring his insistence on cycles, scientists made progress central to the development of the universe.

To many Victorians, progress seemed as inevitable as a Law of Nature. An original chaos of swirling clouds had condensed to form the solar system, a molten ball had cooled into a habitable Earth, and primitive types of life had eventually led to sophisticated organisms. In this age of exploding confidence, improvement was predicted on many fronts—the British empire would expand, machines would become more efficient, education would improve, travel would get faster, and books would get cheaper. Although Charles Darwin's version of evolution denied the possibility that God was directing nature in any particular direction, it did fit in with the concept of progressive development from tiny organisms upwards to human beings.

The foundations of this faith in progress were laid in the eighteenth century, when Erasmus Darwin formulated his own evolutionary theories. Although often called the Age of Reason, this period saw the birth of the consumer society in Britain. The population was increasing rapidly, the rate of inventions was accelerating, and the British economy was booming. Darwin's friend Wedgwood gave his name to fine china, but perhaps his greatest legacy was using advertisements to convince ordinary people they should keep up with the Joneses. Their lives would improve, he promised, if they spent more money on fashionable clothes and the latest line in fancy pottery.

Attitudes towards God were also altering. As a rough guide, in the mid-seventeenth century, Christians felt themselves to be in a dependent relationship: if they failed to follow the laws of God as laid down in the Bible, then He would punish them for their sins. Gradually, there was a turn towards self-monitoring. As Alexander Pope put it in his *Essay on Man* (1711), 'Know then thyself, presume not God to scan/The proper study of mankind is man', that approach to morality reproduced on countless title pages in the eighteenth century.[5] By the end of the eighteenth century, Erasmus Darwin had articulated a further shift, relegating God to the sidelines in daily affairs and making human beings the active agents of their own progress. This was a different position from that said to be adopted by French philosophers. Whereas they were accused of converting men into soulless machines, Darwin's humans were driven by their inner energies.

Europeans used the capacity for self-improvement as a marker of civilization, regarding themselves as being above other peoples, who in their turn were superior to animals. This picture of progress was especially prevalent in Scotland, where philosophers and historians explained that societies had gradually developed from their original barbaric state: this was the 'rudeness to refinement' model of human culture. Influential authors such as the philosopher David Hume and the economist Adam Smith lent their support to the possibility of studying history scientifically. Just as Newton had brought mathematical order to the cosmos, so too, they insisted, rules could be established for demonstrating the progress of humanity. Primitive peoples (their vocabulary, not mine) were preoccupied with finding food, keeping warm, and defending themselves against their enemies. It was only when material conditions improved that they gradually found time to stabilize their systems of government and engage in intellectual or artistic activities.

When travellers reported back on societies overseas, Europeans found evidence to confirm the natural laws of development they were trying to formulate. Although progress had been irregular, commented Gibbon, human beings had gradually emerged from savagery 'to command the animals, to fertilise the earth, to traverse the ocean, and to measure the heavens'.[6] This confidence was shared by the members of the Lunar Society. As they saw it, inventions such as balloons, steam engines, and electricity promised ever tighter control over nature, still greater supplies of the comfort and leisure essential for cultural progress. Priestley stressed the importance of scientific research for ensuring that this optimism was justified: 'it is nothing but a superior knowledge of the laws of nature,' he wrote, 'that gives Europeans the advantages they have over the Hottentots...science advancing, as it does, it may be taken for granted, that mankind some centuries hence will be as much superior to us...as we are now to the Hottentots'.[7]

Britain's expansion in commerce, industry, and empire were indisputable, but not everybody agreed that life was getting better. One problem was deciding what 'better' might mean. Advocates of advance pointed to inventions such as printing, magnetic compasses, and gunpowder that had not existed in the time of the Greeks. Surely, they argued, these represented progress? But their opponents had an easy retort: technological innovations do not necessarily imply moral superiority. On the contrary, they maintained, convenience and wealth encourage decadence, and they praised the Pacific islands as isolated earthly paradises that had not yet been corrupted. For the Lunar men, industrialization was unquestionably

beneficial because it increased profit and efficiency. For their critics, mech-
anization and commercialization heralded an inevitable slide down into
the corroding comfort of luxury.

These warnings about the dangers of excessive consumption were gradually eclipsed by eulogies of progress, but revived during periods of uncertainty. In the wake of the French Revolution, the *Anti-Jacobin* propagandists
challenged the value and meaning of progress. Threatened by an unstable
political situation, they sought safety in maintaining the status quo rather
than venturing into an uncertain—and possibly Frenchified—future.
Searching for an easy target, they picked out an ideal candidate: Darwin,
the radical doctor who had written that the Earth's animals 'have constantly improved, and are still in a state of progressive improvement', the
poet whose old-fashioned couplets and heavy-handed didacticism lent
themselves to parody.[8]

Darwin was the ostensible victim, but Godwin represented more serious prey. In their introduction, the satirists claimed—or rather, they
claimed that Godwin claimed—that 'we have risen from a level with the
cabbages of the field to our present comparatively intelligent and dignified
state of existence, by the mere exertion of our own *energies*'. Sounding like
a pseudo-mathematical Priestley, they imagined an ever-escalating
improvement that 'would in time raise Man from his present biped state
to a rank more worthy of his endowments and aspirations; to a rank in
which he would be, as it were, *all* MIND'. Pretending to speak from a radical vantage-point, they maintained that the only obstacles to such progress
are 'KING-CRAFT and PRIEST-CRAFT, and the other evils incident to what
is called Civilized Society'.[9]

We are, the *Anti-Jacobin* satirists wrote, 'more satisfied with *things as
they are*' than Darwin, but 'less convinced of the practical influence of
Didactic Poems'. Yet their mock-didactic poem did have an enormous
influence on Darwin. He had planned to call his next major book *The
Progress of Society*, but the adverse publicity generated by the *Anti-Jacobin*
encouraged him to change his mind. Eager to dissociate himself from
Payne Knight's *Progress of Civil Society* as well as from the *Anti-Jacobin*'s
'The Progress of Man', Darwin revised his manuscript and gave it a new
title: *The Temple of Nature*.[10]

Both Knight and Darwin were inspired by the same source—Lucretius' *On the Nature of Things*. In Lucretius' godless cosmos, nothing exists except for atoms whirling through empty space, accidentally colliding, rebounding, and combining. Chance rules throughout. The universe has no preordained destiny, no certain future—human beings can pray all they like, but the only strategy that works is to behave rationally. These two Enlightenment poets drew especially on Book V, in which Lucretius reviews the Earth's history, starting from its physical formation, advancing through the origin and development of life, and then on to the establishment of human society.

The full title of Darwin's last and most controversial long poem was *The Temple of Nature; or, The Origin of Society*. It appeared in 1803, the year after he died—and half a century before another book on evolution with a similar-sounding title: *On the Origin of Species*.

CHAPTER TEN

Defining People

[I]mperfection is in some sort essential to all that we know of life. It is the sign of life in a mortal body, that is to say, of a state of progress and change. Nothing that lives is, or can be, rigidly perfect; part of it is decaying, part nascent....All things are literally better, lovelier, and more beloved for the imperfections which have been divinely appointed, that the law of human life may be Effort, and the law of human judgment, Mercy.

John Ruskin, *The Stones of Venice*, II (1853)

Taking refuge from a shower, I once found myself in a bizarre tourist trap specializing in polychrome wizards and illuminated dragons. Piled up on the floor was a collection of...well, what everybody used to call gollywogs, those leftovers from the Victorian era. I tried to justify their presence as antiques on sale for their curiosity value, but they were brand new, shiny smart in their bright nylon jackets and striped trousers. I had spent the last few weeks immersed in old travel narratives, trying not to flinch at terms I found abhorrent, worrying about how to discuss race without inadvertently causing offence. How could anyone think it was acceptable—let alone desirable—to sell these stuffed dolls?

On the other hand, too much political correctness can lead to absurd decisions. Staring at those red mouths sewn into flat black faces, I remembered criticizing a Paris museum that had airbrushed out a cigarette from a photograph of Jean-Paul Sartre. Just as all left-wing intellectuals smoked fifty years ago, so too, even the most enlightened of eighteenth-century abolitionists used names for black Africans that are now offensive. For them, 'negro' was not a term of abuse, but a label specifying western Africans; they were regarded as superior to 'Hottentots', a derogatory label crudely imitating the Khoikhoi language that lumped together a great variety of peoples.

Misunderstandings have bedevilled discussions of race for a long time. In his influential treatise *Spirit of the Laws* (1748), the French philosopher Montesquieu argued that slavery was incompatible with political liberty. Naively assuming that his readers would recognize satire when they saw it, Montesquieu devised a series of absurd consequences that proceed logically from the assumption of black inferiority, but are only funny if you realize that they are self-evidently ridiculous—sugar would be too expensive without slaves to cultivate the plantations, Africans are so stupid they prefer glass beads to gold necklaces, a wise God would not place a clean soul in an ugly black body. And so on. Unfortunately, the jokes misfired, and Montesquieu has often been cited as an exemplar of bigotry.[11]

As a research project proceeds, you often forget your initial state of ignorance. When I started, I was unaware how misleading it is for books about race to bracket together the decades immediately before and after 1800. I had inherited the biological distinctions of race that seemed so obvious to imperialists, but I came to realize that these were unknown to Enlightenment explorers. It was only in the nineteenth century that people with similar physical characteristics, rather than similar origins, were definitively grouped into separate races. Race started to be a scientific term, a measure of distinction based on supposedly quantifiable factors such as skin colour and skull shape.

Another false assumption is that the same debates were taking place everywhere. Many definitions of race originated in German universities, where scholars had little direct experience of slavery. Why should their views be shared or even known about by British abolitionists or Caribbean plantation owners? 'Caucasian' is now entrenched as a fundamental human category, but its first use in English seems to have been in 1807, almost thirty years after the anatomist Johann Blumenbach introduced it in Germany. Oddly, a mountainous region bordering Europe and Asia, remote from scholarly centres, has given its name to a major racial group.[12]

The notion that humans are biologically divided might seem a scientific innovation, but it was promoted by economic concerns of the time and carried political implications for the future. Debates about democracy, race and slavery were intertwined. It served the interests of American planters to insist that their slaves belonged to an inferior race and hence (according to their logic) could be treated as if they were mere possessions, material objects rather than sentient beings. As Jeremy Bentham put it in his recommendations for legal reform, animals 'stand degraded with the class of *things*' and slaves are treated 'like inferior races of animals'.[13] In Europe after the French Revolution, rich aristocrats warded off the threat of equality by insisting that innate differences are scientifically proven. Even Benjamin Franklin, self-proclaimed champion of republicanism, was so concerned about Pennsylvania being inundated with 'swarthy' German immigrants that he declared them inferior to English Saxons.[14]

The abolition movement interacted closely with other anti-cruelty campaigns, and prompted discussions of human/animal distinctions. Plantation owners and their allies relied mainly on economic rather than racial justifications, maintaining that slaves could be purchased like animals and were essential for maintaining profitable trade. In contrast, abolitionists classified slaves as human beings and proposed various theories about differences between Africans and Europeans. Their insistence on discussing such issues affected scientific theories of human origins and environmental influences.[15]

The more I thought about Erasmus Darwin and abolition, the stranger it seemed to me that relatively little attention has been paid to the early arguments about human differences that were being conducted during his lifetime. The nineteenth-century shift towards anthropological taxonomy had such serious long-term consequences—Nazism, apartheid—that its roots need to be explored thoroughly.

Eighteenth-Century Britain

Writing in eighteenth-century Britain about the human races is difficult because anthropologists had not yet invented them: the word 'race' existed, but its meaning was different from the modern one. Race was defined by social relationships, not by physical features. Although European explorers regarded the indigenous inhabitants of America, Africa, and Australia as being different from themselves and from each other, they included them in the same race: they were all human beings, as opposed to dogs or horses. Sometimes the French and the British were said to be two races because they lived in different countries; extended families—what we might call clans—could also be termed races. Towards the end of the century, 'Indo-European' peoples were bracketed together into a single race by followers of William Jones, who stressed their common linguistic roots. What mattered was a shared origin, not a bodily resemblance.[16]

Cultural prejudices and preconceptions abounded, but they were not the same as current ones. Although Jews were excluded from the universities, this was on grounds of religion: like Dissenters and Catholics, they were discriminated against because they were non-Anglicans, not because they belonged to any supposed international Semitic community. The distinctions between Black and White were far from black and white. The terms were adjectives rather than nouns, and classifications were mutable, not fixed. Many doctors believed that Europeans acquired dark skins—not just suntans—when they lived in equatorial climates. France's most celebrated naturalist, Georges-Louis Leclerc, Comte de Buffon, reported as fact that an adopted African baby turned white when it was brought up in Holland, and the abolitionist Thomas Clarkson helped to spread the belief that when children of African slaves are born in a cooler climate, they are more light-skinned than their parents.

In their own minds, British gentlemen sat confidently at the peak of the human hierarchy, but the ranking of those lower down the scale—women, non-Europeans—was still fluid. 'Is one half of the human species,' protested Mary Wollstonecraft, 'like the poor African slaves, to be subjected to prejudices which brutalize them…?'[17] After an aristocrat was spotted flirting in a garden with her black servant, gossip columnists reported that 'The ladies have got a new phrase for infidelity in *shrubberies*—they call it being *transported* to the *plantations*' and recommended installing appropriate hardware of '*Steel traps and spring-guns*'.[18] On the other hand, female slaves working in the American plantations were often praised as being submissive and maternal—precisely those characteristics that London husbands were encouraging in their wives and daughters.[19]

Slavery is a term often reserved for black people of African origins, but in a broader sense it had always existed all over the world—and it still does. European traders started to buy African slaves not because they were seen as racially inferior, but because they were cheaper and more abundant than those available from other sources. According to the ethical codes of the time, for one Christian to subjugate another seemed immoral, and a first step taken by many escapees was to become baptized and so gain protection against being forced back into service. One English traveller described Gambians as 'Black as Coal'—but reported that 'thro' Custom, (being Christians) they account themselves White Men'.[20]

When explorers published accounts of their adventures overseas, they were often more interested in attracting readers than in telling the unadorned truth. However, men like Erasmus Darwin who had never been abroad had little else to go by. Although highly romanticized, these subjective visions of distant continents were the only ones available, so their influence was pervasive and long-lasting. I discovered a letter to Wedgwood in which Darwin recommended boosting the abolitionists' cause by reprinting an abridged section from a novel that had been published decades earlier—*Colonel Jack* (1722), an imaginary travelogue by Daniel Defoe about a young English man who finds himself kidnapped to become a slave in Virginia. Sitting in the library, deciphering Darwin's script, I laughed (silently), because Darwin's verdict on Defoe was precisely the criticism often levelled at him: the story was, he told Wedgwood, 'told rather too diffusely, so as to be almost tedious'.[21]

I had never heard of *Colonel Jack*, but a short time later a copy was on the desk in front of me. Published three years after *Robinson Crusoe*, the book describes a reformed pickpocket who ingratiates himself with his master and becomes a wealthy plantation owner. When Jack is first promoted from slave to overseer, he loses control over his charges because he is reluctant to administer the painful whippings that he has himself so recently experienced. Eventually, he concedes the possibility that Africans may be capable of human emotions, and devises an insidious form of psychological manipulation relying on the threat rather than the actuality of physical cruelty. First Jack terrifies a slave with descriptions of the agonizing punishment lying ahead, and then at the last moment apparently switches his allegiance, dispensing mercy so that the victim feels bound into subservience by a debt of gratitude. As Jack explains to his master, 'your Business shall be better discharg'd, and your Plantations better order'd, and more work done by the Negroes, who shall be engaged by Mercy and lenity, than by those, who are driven, and dragg'd by the Whips, and the Chains of a merciless Tormentor'.[22]

Darwin's letter to Wedgwood still lay on my desk, and on a second reading I interpreted it differently. He singled out for praise Defoe's description 'of the generous spirit of black slaves—when kindly used'. Like Jack, Darwin was suggesting not that it is immoral to torture slaves, but that you get more work out of them when you treat them well. It seemed that although Darwin was genuinely horrified by the vicious punishment meted out on the plantations and the transport ships, he also subscribed to the ingrained assumption of his peers that wealthy gentlemen deserve to benefit from a cheap energetic labour force.

Returning to Defoe, I discovered that when Jack himself becomes a rich planter, he initially remains 'Grateful to the last degree to my old Master, who had rais'd me from my low Condition and that I lov'd the very name of him, or as might be said, the very Ground he trod on'.[23] Although this insight leads Jack to a new faith in God, he persists in regarding his black slaves as intrinsically inferior and so incapable of transferring their own adoration from a human to a divine ruler.

This trope of a grateful slave, the honest African who devotedly serves his cruel master, came to permeate Anglo-American culture. I looked again at Wedgwood's medallion (see Fig. 7, p. 180), and saw it in a new light. In my reinterpretation, the kneeling African is praying for sympathy, the sympathy that he has been taught to expect from his white masters if he behaves obsequiously. His submissive pose offers the reassuring message that liberated Africans will not abuse their new-found freedom by rebelling, but will display their gratitude through continued subservience. This slave is demanding to be well treated and to have his chains removed because he is a fellow human being—but he is not claiming the right to equality.[24]

Origins

Racial classification originated in the eighteenth century, dominated anthropology in the next, and precipitated mass slaughters during the one after that. Racial discrimination is illegal in modern Britain, but the assumption still survives that there are real distinctions between human races. However, there are no cut-and-dried criteria about where the lines should be drawn.[25]

All debates about classifying the natural world are riddled with subjectivity, and many of them become acrimonious. When Darwin wrote *The Loves of the Plants*, he knew that he was defending Linnaeus against some powerful critics, most importantly Buffon, who accused him of arbitrarily relying on one single characteristic, a flower's sexual organs, to divide up God's creation into artificial groups. The universe is made up of individuals, insisted Buffon, so that any classification system is merely an imaginary one drawn up for convenience.

As well as disagreeing about plants, Linnaeus and Buffon had very different views about classifying people, and it was in the course of their arguments that the word 'race' started to be used with something approaching its modern meaning. The questions were clear, but not their answers. Do all human beings originate from the same source, or were there several independent creations? Should living creatures be sorted into separate groups, or should they be organized into a single hierarchical scheme? Are characteristics fixed, or can they be fashioned by factors such as climate, education, and occupation—and then inherited? What is the relationship between human beings and animals?

For much of the eighteenth century, many French and British people believed that all human beings belong to one species. As written in the Bible, God 'hath made of one blood all nations of men for to dwell on all the face of the earth'.[26] Although the word had not yet been invented, this view later became known as 'monogenism', meaning 'of a single origin', which corresponds to the story in Genesis that all human beings are descended from God's first couple, Adam and Eve. Adam and Eve were assumed to be white, and black people were said to have degenerated from this original, natural colour.

In contrast, according to 'polygenists' (a nineteenth-century word), there were several independent creations, so that human beings vary not because they have deteriorated over the generations, but because they have different ancestors. Polygenism was more common in America than in

Britain, perhaps because plantation owners felt more comfortable mal-treating a biologically distant relative than a close one (after all, they argued, nobody thought there was anything wrong in chaining up an animal). Ostensibly, multiple origins contradict the biblical account of creation, and there were two main ways of reconciling the clash: to claim that people must have existed before God created Adam (pre-Adamism), and to place the blame on Ham, third son of Noah and directly descended from Adam and Eve.

The most recent version of pre-Adamism, I discovered, had been invented to resolve a logical conundrum that perplexed literalist theologians (very tempting to mutter that some people have nothing better to do with their time). If human law was established only after Adam had sinned in the Garden of Eden, there must have been a lawless people who existed before him—and so how could he be the first man? The solution was ingenious: since Adam was the father of the Jews, then the Gentiles had been created separately and earlier. This version offered an additional advantage for biblical exegetes, because it brought into existence a non-Jewish wife for Adam's son Cain to marry after he had killed his brother Abel. The descendants of this evil murderer and an inferior woman were doomed in advance, cursed forever by the mark of Cain, often said to be black.

The other solution was to suggest that human beings had split into three races immediately after the great Flood, which Noah survived by building an ark for his family and representatives of all the animals. According to Genesis, once the waters had receded, God blessed Noah and his three sons, telling them to go out and populate the world. All went well until Noah cultivated a vineyard, got drunk on his own wine and fell asleep naked in his tent. When Ham spotted Noah, he invited his older brothers to share the view, but they virtuously covered up their father without looking at his body. On some interpretations, as soon as Noah woke up, he cursed Ham's son Canaan (why Canaan should take the rap remains mysterious), who was consigned to a life of servitude and whose offspring were black. (This tidy explanation encountered some obstacles after Europeans found the Americas. Since Noah's ark landed in the Old World, how did the indigenous Indians manage to reach this distant continent and survive? Swimming was clearly out of the question.

However extraordinary it may seem that such questions could be asked, asked they were.)

Few eighteenth-century polygenists were brave enough to go into print and challenge the Christian convention of a single human creation. One who did was a naval surgeon called John Atkins, who believed that 'the black and white Race have, *ab origine*, sprung from different-coloured first Parents'; sensibly protecting himself, Atkins admitted that this view was 'a little Heterodox'.[27] Two other proponents were far more famous—the philosopher David Hume and the distinguished Edinburgh judge Lord Kames (confusingly, also referred to as Henry Home). Although their arguments are now often interpreted as being about race, they were launched in the context of theological debates.

Because Hume was interested in attacking the traditional account of creation in the Bible, he emphasized differences. There must, he wrote, have been 'an original distinction between these breeds of men' (interesting agricultural term to choose), and so he had come 'to suspect the negroes to be naturally inferior to the whites'.[28] In contrast, Kames wanted to defend scriptural testimony, and he started by observing (or claimed he did—he had never crossed the Atlantic) that American Indians all looked the same as each other but were distinct from Europeans. Ingeniously, he suggested that the New World was indeed new—that it had arisen from the ocean long after the original Creation. This left Adam and Eve free to populate the Old World with their progeny, while God could plant a new race in America.[29]

The most notorious polygenist was Edward Long, a Jamaican planter whose invective has made him a favourite source for historians in search of juicy quotations. His *History of Jamaica* (1774)—three hefty volumes—is a pseudo-philosophical package designed to justify not only owning slaves, but also maltreating them. Peppering his text with emotional words—'brutish', 'bestial'—Long informs his readers that black Africans are subhuman: they devour human flesh, tear at raw meat with their talons, are infested with black lice, smell noxious, leave their children to run wild, and are incapable of logical thought. Living like animals, they become expert swimmers who can stay below the water an inordinately long time.[30]

Long's fellow plantation owners may have appreciated learning that slaves could be regarded as animals, but after much reading, it seems to me that in eighteenth-century Britain his views were probably as unrepresentative as those of the British National Party today. Gazing at Long's diatribe in the Rare Books Room, I wondered why it is still quoted so frequently and at such salacious length. Is it because by making the past appear primitive, modern historians can preach a fashionable mantra of progress?

I explored my own feelings. Would I become inured to Long's invective if I went on studying it? Should books like this be destroyed? Did Long really believe that Africans have sex with apes? I remembered the early panic surrounding AIDS, when the same claim was made to explain the origins of the HIV virus. Examining such unpalatable evidence forces us to confront the reality that, in some ways, there has been no progress.

Permanence or Change?

As all good parents know, order and stability are extremely reassuring. Georgian paternalists intended the sense of stately symmetry that characterized early eighteenth-century music and country mansions to reflect the harmonious organization of God's world. As theologians and naturalists struggled to make sense of the cosmos, they imposed a neat hierarchical structure in which every being is ordained its permanent settled place: the Great Chain of Being, a Christianized version of Aristotle. In this ladder-like structure, creatures are arranged progressively with only minute differences between neighbours. Starting with the tiniest animalcules, worms, birds, animals, and then people are ranked in succession before ascending yet further towards angels, and finally God Himself. Pope described the Chain in his *Essay on Man*, the most influential English literary work of the Enlightenment:

> Far as Creation's ample range extends,
> The scale of sensual, mental pow'rs ascends:
> Mark how it mounts, to Man's imperial race,
> From the green myriads in the peopled grass...[31]

For anybody who favoured the status quo, this hierarchical organization provided a convenient way of justifying inequality. After all, if God had set this scheme in place, then surely mere mortals should not tamper with it? Soame Jenyns, a wealthy landowner, used the Chain to justify treating Africans as inferiors:

> animal life rises from this low beginning in the shell-fish, through innumerable species of insects, fishes, birds, and beasts to the confines of reason, where, in the dog, the monkey, and chimpanzee it unites so closely with the lowest degree of that quality in man, that they cannot easily be distinguished from each other. From this lowest degree in the brutal Hottentot, reason, with the assistance of learning and science, advances, through the various stages of human understanding, which rise above each other, till in a Bacon or a Newton it attains the summit.[32]

But there were other ways of interpreting the Chain to suit an argument. In America, when Thomas Jefferson—plantation owner as well as champion of equality—suggested that black Africans belong to a different race from white Europeans, an abolitionist accused him of contravening divine intentions: 'You have degraded the blacks from the rank which God hath given them in the scale of being!', he protested; 'You have advanced the strongest argument for their state of slavery!'[33] Darwin railed against the Chain by criticizing its implicit vindication of suffering: 'Pray ask... if this pain is necessary to establish the subordination of different links in the chain of animation? If one was to be weaker and less perfect than another, must he therefore have pain as a part of his portion?'[34]

The Chain exploded and then collapsed as international explorers brought back increasing numbers of exotic specimens that would not fit into its neat scheme. As one alternative, Linnaeus suggested dividing plants and animals into several large groups, each repeatedly subdivided into smaller and smaller ones. Linnaeus attributed great significance to the number four, redolent not only with mystical but also with biblical symbolism, such as the four rivers dividing the Garden of Eden. He identified four human colour-groups, each corresponding to a continent and to one of the Aristotelian temperaments then still prevalent in medicine: the red choleric Americans (known as Red Indians for the next couple of centuries), the white sanguine Europeans (the only *homo* to be honoured with the label *sapiens*, or 'wise'), the yellow melancholy Asians, and the black phlegmatic Africans.

This basic concept proved extremely influential, and formed the basis for important racial groupings in the following century. However, Linnaeus avoided the theological Big Question: how could one single original couple, Adam and Eve, have generated descendants who looked so different from each other? There were two main lines of argument, one based on the means of survival and one on climate.

To survive, let alone to flourish, people need food, shelter, and clothing. Theories based on the mode of subsistence originated with the English political philosopher, John Locke, and became especially important in Scotland. The details varied, but essentially there were four stages, which represented successive states of both material and moral improvement. Originally, people had lived by hunting animals, but they gradually advanced to a more secure type of existence based on nomadic farming. The third stage featured settled agriculture, while the final plateau (Europe, naturally) was inhabited by stable societies linked to one another in commercial trading networks.

This theory of development could be used to justify oppression because incipient hierarchy was built into it. Hunter-gatherers operated in small groups, but when communities became established, laws were devised to maintain order—which meant that some people imposed restrictions on others. In more advanced societies, ran the argument, privileged members became employers while others served or became enslaved; elites emerged who read, studied, and enjoyed music, while lesser beings endured a more menial existence. Impoverished labourers or nomadic Africans could hardly be expected to share the sophistication of wealthy gentlemen running large estates. For Scottish philosophers such as Adam Smith and David Hume, European cities—notably Edinburgh—represented the apogee of civilized culture.

The other important type of model was based on climate. After the Fall, when Adam and Eve were banished from the Garden of Eden, people spread to the four corners of the Earth, where their environment affected them in different ways. This central concept of geographical adaptation was old, dating back to Hippocrates and Herodotus, but during the eighteenth century it reappeared in various versions, all of them relying on the same basic arguments—Africans were permanently blackened by the sun, and became lazy because food is so easily available, whereas freezing conditions in the Arctic region stunted the inhabitants' physical and mental development. In contrast with both those extremes, Europe lies in a temperate zone, and so provides the ideal environment for encouraging the refined civilization that stems from a combination of work with leisure.

Climate theory lent itself to the rationalization of prejudice. By far its most influential exponent was Montesquieu, who cited France (obviously) as the ideal place to live, with neither the extreme heat fostering indolent impulsiveness nor the icy cold inducing savagery. An eloquent advocate of democracy, Montesquieu outlined an environmental theory of politics—but one with a clear bias. Because Europeans all live in equable environments, he maintained, their countries are well matched and can coexist in equilibrium. Conversely, Asia's nations, which extend across several climate zones, are unequally balanced: the strong overpower the weak, encouraging despotism and slavery.

A further twist to climate theories was provided by Buffon. Like Linnaeus, he recognized that the Chain of Being had to change, but he proposed a very different solution. Buffon brought in time—lots of time, far, far more than the six thousand years calculated by biblical literalists. Just as controversially, he suggested that life on Earth had changed since the Creation. Instead of being static, argued Buffon, the Chain represented a long slow deterioration from the original perfection of human beings in the Garden of Eden. Over many millennia, living creatures have degenerated to produce the current enormous variety that ranges from white Europeans at the top down to the lowliest creatures.

As France's most eminent naturalist, Buffon was well placed to persuade doubters. His version of the animal kingdom was far more than just another modification to existing thought: by introducing an extended past of gradual change, it drastically revised ingrained assumptions about the world and its inhabitants. Supporting Buffon's ideas, climatic principles explained why people had deteriorated as they dispersed round the world. For instance, it made sense (then) to argue that Arctic conditions caused the

Sami (Lapps) to be stupid, ugly, and superstitious; conversely, British abolitionists cited the affidavit of an American doctor: 'The Spaniards, who have
inhabited America, under the torrid zone...are become as dark coloured as
our native Indians of Virginia; of which I *myself have been a witness.*'[35] Influenced by Buffon, Darwin—a teetotaller who had watched his first wife die
through alcoholism—warned readers against marrying an heiress, in case
she should prove to be the last member of a degenerating family.[36]

When 'The Loves of the Triangles' appeared, the satirists' fall guy—
Darwin—was far from alone in suggesting that some sort of evolutionary
development had taken place. Fossils and other geological evidence had
already convinced many naturalists to push back the date of the Creation
and propose a universe that was constantly changing. Darwin's ideas about
progress were similar to those being developed by the Parisian biologist
Jean-Baptiste Lamarck, who first published them slightly later, in the
opening decade of the nineteenth century. Along with many of their contemporaries, Darwin and Lamarck thought that characteristics acquired
during people's lives—a blacksmith's muscles, for example—can be passed
on to their children.

Fifty years after that, Darwin's grandson Charles eventually released *On
the Origin of Species*, the book often said to have revolutionized human
thought by introducing evolution. But by then, the concept of gradual
development over aeons of time had long ceased to be revolutionary.
Charles Darwin's innovation was not to suggest evolution, but to outline
a theory of how it might happen—natural selection. Even so, unable to
provide a complete explanation, he argued that acquired characteristics
can be inherited. Rather than overturning the beliefs of his grandparents'
generation, Charles Darwin was in some ways reviving them.

Animals/Humans

Sometimes carelessness pays dividends. One afternoon, I pressed the return
key on my computer too quickly, and instead of Mary Wollstonecraft's
A Vindication of the Rights of Women, found myself confronted by the title
of another tract that came out in the same year—*A Vindication of the Rights
of Brutes* (1792). Its author was Thomas Taylor, a one-time bank clerk who

had retired on the profits of his phosphorus lamp (allegedly perpetual), but was more famous for translating Aristotle and Plato. This parody of Wollstonecraft was Taylor's sole venture into contemporary debates.

To my surprise, I discovered that Taylor and his wife had temporarily taken in Mary Wollstonecraft while she was a rebellious, unhappy teenager. Presumably relationships deteriorated after that, since Taylor's satire on her ideas seems vindictive. Although human beings may have superior powers of reason, he argued, lions are stronger, spiders can spin webs, and hares can outrace us. How far down the Chain of Being should we draw the line, he asked with mock ingenuousness? Should vegetables and minerals also be awarded rights? By relentlessly spelling out the consequences of extending rights to creatures lower down the scale, Taylor reduced Wollstonecraft's arguments to a logical yet absurd conclusion. And not only hers—Taylor was, like Wollstonecraft, responding to Thomas Paine's *The Rights of Man*, his defence of the French Revolution that sold thousands of copies despite being officially banned.

Like many other Enlightenment men, Taylor was more interested in securing his own moral welfare than in altruistically alleviating the condition of others. Because God has placed human beings at the top of the Chain, ran the argument, the virtuous and the civilized are distinguished by treating those lower down with compassion. For them to behave otherwise would be to demean themselves: cruelty affects the perpetrator as well as the victim. This was the same rationale that governed men's attitudes towards women and slaves. As Bentham put it, 'The question is not, can they *reason*? Nor, can they *talk*? But, can they *suffer*?' Torturing animals, beating wives or shackling slaves was deemed wrong because it was inhumane, not because it contravened rights.[37]

For some plantation owners, treating their slaves like animals seemed justified: after all, if slaves were merely beasts who resembled people, then locking them up was hardly maltreatment. Since God had placed humans at the summit of creation to act as the world's custodians, it was a Christian's duty to treat lower creatures with kindness, and campaigners for abolition were also involved in animal welfare. In 1824, William Wilberforce helped to found the world's first charity to protect animals, and offenders were successfully brought before the courts. Fifty years went by before a similar society was established in New York to rescue abused children. When British philanthropists followed suit, they adopted what now seems an extraordinary argument—since babies are little animals, they too are covered by legislation and should be taken into care.

Witty (but not subtle), Taylor's satire worked because his readers were convinced that however successful the Revolution, one barricade would never tumble: the distinction between animals and people, who had been created separately by God on different days. Animals might remember, show affection, or learn tricks—yet, as Pope maintained, they could never become human:

> Remembrance and reflection, how allied:
> What thin partitions sense from thought divide;
> And middle natures, how they long to join,
> Yet never pass the insuperable line![38]

Some authors did dare to probe the human–animal boundary, and one particularly easy target for the *Anti-Jacobin* satirists was Lord Monboddo (aka James Burnett), to whom Darwin referred approvingly.[39] An eminent Scottish judge, Monboddo cultivated his reputation for eccentricity. Convinced that civilization causes degeneration, he apparently exercised naked, delighted in cold baths, and rode his horse to London dressed in a scarlet cloak, reading Homer while his black servant Goro followed a few paces behind. Crank he may have been, but he moved in influential circles. An intellectual gourmet, Monboddo threw dinner parties that attracted Samuel Johnson, James Boswell, and Edinburgh glitterati such as Robert Burns (who fell in love with Monboddo's daughter).[40]

Perhaps because he was deaf, Monboddo focused on language as the major marker of human progress. According to him, animals with vocal organs similar to those of humans must belong to the same species. He had never seen any orang-utans, let alone visited Asia to study them in the wild; in any case, the small number said to have reached Britain were actually chimpanzees. Even so, Monboddo insisted that the bodies of orang-utans are essentially the same as those of people. Surprisingly gullible for a judge, Monboddo took at face value not only the dissection of a chimpanzee performed almost a hundred years earlier, but also travellers' reports that orang-utans play the harp, use sticks for weapons and carry women off, King Kong style.

Having shown (he thought) that orang-utans are fundamentally the same as people, Monboddo next approached his topic from the opposite direction. Crucially, he believed that language is not an intrinsic capacity, but an acquired skill. To support this premise, he pointed out that neither babies nor mutes can speak, and yet they are undoubtedly human. Monboddo also had another strong source of evidence—children who had grown up in the woods, deprived of contact with other people. For this part of his argument, Monboddo inspected for himself two of Europe's most famous wild children: Peter of Hanover, and the feral girl of Champagne, Marie-Angélique Leblanc.

Peter had been captured in a German forest as a naked young teenager rooting around for acorns and berries. Cleaned up and dressed, he was summoned to the English Hanoverian court, and briefly became a curiosity until the novelty waned and he was dispatched to live with a farmer in Hertfordshire. Peter was in his late sixties when Monboddo visited him, still unable to say more than a few words, but polite and well-behaved. Leblanc presented a very different case. Reportedly an athletic Sioux Indian who had thrived on raw meat and fruit, her physical health deteriorated in captivity, but she became fluent in French, moved into a convent and somehow (this part of her life remains as mysterious as its beginning) ended up as a wealthy society hostess in Paris.[41]

Like Godwin, Monboddo believed in perfectibility, and he used Peter and Leblanc to illustrate his theory of progressive improvement. All members of our species—orang-utans as well as humans—are born capable of learning language, but need to be taught it. Peter remained at the infantile stage, able to stand erect and understand words. Orangutans have developed further, because as well as walking on two feet, they communicate with each other and show evidence of social organization. Finally, Leblanc, who could swim and talk, represented a higher level intermediate between apes and humans. If a few bright orang-utans could be trained up at the Edinburgh school for the deaf, suggested Monboddo, they would soon develop their innate talents and speak Scottish.[42]

Unsurprisingly, Monboddo came in for a good deal of ridicule: in France, the anecdote circulated that a cardinal had told a caged orang-utan: 'Speak, and I will baptize you.'[43] Monboddo made himself delightfully easy to mock by maintaining that wild men used to have tails before they became civilized, and the *Anti-Jacobin* satirists took advantage with panache. 'Others by an inherent disposition to society and civilization, and by a stronger effort of *volition*, would become Men,' they wrote in a footnote to 'The Loves of the Triangles'. 'These, in time, would restrict themselves to the use of their *hind feet:* their *tails* would gradually rub off by sitting in their caves or huts, as soon as they arrived at a domesticated state: they would invent *language*, and the use of *fire*, with our present and hitherto imperfect system of *Society.*'[44]

On the other hand, decades later, Monboddo was presented as a precursor of Charles Darwin. In the opinion of a Victorian judge:

> Though Darwin now proclaims the law
> And spreads it far abroad, O!
> The man that first the secret saw
> Was honest old Monboddo.[45]

Continuities

'Segregation now, segregation tomorrow and segregation forever!'[46] Following his 1963 inaugural address as governor of Alabama, George Wallace gained an alarmingly high number of votes in his campaign to become president of the United States. Attempting to make prejudice appear rational, Wallace claimed that race is a natural category and so provides a valid justification for sending children to separate schools. He was wrong: there is no biologically clear-cut or objective way of classifying living beings. Rules have to be invented and then agreed on. Is a bat a bird because it has wings, or a mammal because its young are born live? Swans are white—but what about those large black birds in Australia? What is the racial type of children with three European or Hutu or Jewish grandparents and one who is African/Tutsi/Aryan? There are no absolute right or wrong answers to such questions.

Deciding how to classify plants and animals may have less momentous consequences than when people are involved, but the principles are the same. Linnaeus broke the Chain of Being, but he made subjective decisions—much criticized by Buffon—about his criteria for defining groups. Although Linnaeus is acclaimed for heading towards the future by making taxonomy more scientific, marshalling human beings into races did not necessarily represent progress. Naturalists who were already prejudiced placed black Africans in a separate race from white Europeans, giving discrimination a purportedly objective basis. The logic had got stuck in a loop. Step 1: blacks are inferior, therefore they belong to a separate race. Step 2: blacks are in a separate race, therefore they are inferior.[47]

Another suggestion made in the name of science also had the unintended consequence of justifying white superiority. Pieter Camper, an influential Dutch anatomist and artist, fiercely opposed attempts to create distinct races. A monogenist, he underlined his abhorrence of slavery by writing: 'In the beginning, a single man had been created by God, to wit Adam, to whom all of us, whatever may be our figure or colour, owe our origin.' Trained at Leiden, one of Europe's leading medical universities, Camper lived for six months in England, where he was a Fellow of the Royal Society and published his conclusions.

For Camper, there was an uncrossable barrier between people and animals: moreover, all people were equal. To explain how they could be fundamentally the same and yet look different, Camper adopted a climatic approach, explaining that whites and blacks could interchange if they lived long enough in the right surroundings. Unfortunately, Camper set about proving his convictions in a way that appeared to demonstrate precisely the opposite of what he believed. Arguing that—unlike skin colour—people's skulls are unaffected by environmental influences, Camper decided to compare their dimensions mathematically. Such a research project only became possible towards the end of the eighteenth century, when overseas expeditions started to collect non-European skulls in sufficiently large numbers. For each one in his sample, Camper measured the angle of slope from its forehead down to the upper teeth, and then displayed the profiles on a grid.

Camper compiled a chart showing sixteen skulls arranged in mathematical sequence from a tailed ape and an orang-utan at the left, through 'Negroes' and 'Calmucks', with a European and a Greek Apollo on the right. When you inspect them closely, these diagrams do indeed confirm Camper's insistence that there is far more similarity between the humans than between them and the primates. The visual message that emerges more strongly is, however, a very different one: that the geometric records of the skulls corroborate a racial hierarchy, a progressive scale from animals to African blacks to Europeans.[48]

It was this distorted interpretation of Camper's theory that had the greater impact. When the eminent surgeon John Hunter gave lectures, he reportedly 'had a number of skulls, which he placed upon a table in a regular series, first shewing the human skull, with its varieties, in the European, the Asiatic, the American, the African; then proceeding to the skull of a monkey, and so on to that of a dog; in order to demonstrate the gradation both in the skulls, and in the upper and lower jaws'.[49] Although Hunter denied abolitionists' accusations that his anatomical theories justified inhumane treatment, his lectures fostered prejudice in others.

In particular, Hunter so impressed Charles White, a Manchester doctor, that he started to give his own talks, eventually publishing an influential book called *An Account of the Regular Gradation in Man, and in Different Animals and Vegetables* (1799). White's drawings looked like Camper's, but the underlying philosophy was radically different. Unlike Camper, White favoured polygenism, and so had no need to rely on climate for explaining variations of colour among human beings: for him, differences ran far more than skin-deep. According to White, Europeans and Africans range continuously not merely in their facial angles, but also in their intellectual powers, skeletal structures (blacks have ape-like long arms) and other physical characteristics—sexual organs (simian again), hair, sweat, muscles, and skin.

White claimed to be against slavery, but his anthropological views offered ideal ammunition for its supporters. Conversely, ingrained attitudes towards slaves affected racial concepts that were said to be established scientifically. Historically, male slaves had often been Africans and Tartars, and since they carried out tasks demanding brute force, they were classed as ugly. In contrast, the luxury end of the female market was supplied from the Black Sea area, so that Caucasian women—captured as sex slaves—became epitomized as ideals of beauty, a judgement endorsed by Immanuel Kant. Europe's most influential anthropologist in the early nineteenth century was Blumenbach, who initially stressed the importance of climate, but in later books increasingly conflated white beauty with Caucasian origins.[50]

White? Caucasian? Neither whiteness nor blackness is inherently definable—and how curious to name a self-defined superior tranche of humanity after a region renowned for supplying women sold into sexual servitude. Originating at the very end of the eighteenth century, for almost two hundred years these labels were used more or less interchangeably, with no explanation or justification.

CHAPTER ELEVEN

The Temple of Nature

I cannot sit idly by in Atlanta and not be concerned about what happens in Birmingham. Injustice anywhere is a threat to justice everywhere. We are caught in an inescapable network of mutuality, tied in a single garment of destiny. Whatever affects one directly, affects all indirectly.

Martin Luther King Jr, *Letter from Birmingham Jail* (1963)

A butterfly's wings flapped, an Icelandic volcano erupted, and the whole world succumbed to its greatest traffic jam ever. Trapped in a picturesque medieval city by a faraway cloud of ash, I started to behave like the philosopher Immanuel Kant. Every day—or so at least, the story goes—Kant paced round his home city of Königsberg (now Kaliningrad), repeatedly crossing the seven bridges that span its two rivers with such regularity that local citizens used his appearance to check the time on their clocks. My daily tour around Girona's bridges was less pre-programmed, but its increasing familiarity left me plenty of time to think about Erasmus Darwin.

Could Kant have crossed each bridge once and only once on each trip? This may sound like an esoteric conundrum fit only for abstract mathematicians with time on their hands, but the solution (no: it's impossible) lies at the heart of the theoretical substructure underlying the World Wide Web. Sitting in an Internet café, I silently paid tribute to Kant and his bridges, remembering how I had felt thirty years earlier when stranded abroad bereft of emails, mobile phones, or cash machines. This time was different: although separated from my books, I could explore the Web and convert an inconvenient stroke of fate into a serendipitous opportunity. Almost immediately, Spanish Google threw up an article in a French

Canadian journal that I had previously missed.[51] Under its influence, I started to think yet again about *The Temple of Nature*, Darwin's most complex and most interesting poem. Every time I revisited it, I felt that I reached a new but still incomplete level of understanding: research is an ongoing process rather than a completed act.

The Temple of Nature, Darwin's versified version of evolution, proved to be his longest and most ambitious poem, although it was commercially the least successful. By the time it appeared, the *Anti-Jacobin* was no longer in existence, but the satirists would have had a field day. Darwin was hardly securing himself a place in the poetic pantheon with couplets like these:

> So the lone Truffle, lodged beneath the earth,
> Shoots from paternal roots the tuberous birth...
> Unknown to sex the pregnant oyster swells,
> And coral-insects build their radiate shells;[52]

Whatever his original intentions may have been, Darwin ended up producing an even more shocking vision of change than Knight. The very title—*The Temple of Nature; or, the Origin of Society*—resonated with pagan overtones and recalled anti-Christian propaganda. Mock temples were scattered around many of England's country estates, where Temples of Venus provided convenient meeting points for secret assignations and Temples of Flora were decorative follies. Sacrilegious French Revolutionaries renamed cathedrals as 'Temples of Reason', and staged a dramatic ceremony inside Notre Dame at which Liberty (actually an off-duty opera singer) performed a traditional ritual freshly minted for the occasion. Adorned in white robes and a Phrygian cap, she bowed to the flame of Reason in the nave before sitting down on a bank of flowers. By the end of the century, radical atheists in London had set up their own Temple of Reason in a London salesroom.[53]

For potential purchasers, Darwin's *Temple of Nature* would have sounded ominously like the *Temple of Flora* (1799) by Robert Thornton. An opportunistic physician, Thornton had tried but failed to cash in on sexualized Linnaean botany by commissioning illustrations that were luxuriously coloured and spectacularly erotic. His venture was a financial flop. Darwin's poem is also a paean to sex, but a very different one: non-pornographic, with a few black-and-white images, it joyously hymns reproduction as a natural process that has evolved over time and guarantees progress into the future. And as I have increasingly come to appreciate, his clumsily expressed verse conceals profound philosophical arguments.

The pictures in Darwin's *Temple of Nature* were designed by Fuseli, and his four high-quality black-and-white engravings exaggerate its sexual exuberance. To accompany the poet's celebration of human creativity, the artist passed over his tributes to Handel's music and Newton's gravity, depicting instead a recumbent maiden stretching out languorously towards some sexy Seraphs who 'hover on enamour'd wing'. In a near-parody of the creation story in Genesis, an ecstatic Eve with flowing golden locks emerges from the groin of an Adam endowed with a six-pack. According to Darwin's words, Eros is as 'Warm as the sun-beam, pure as driven snows'; accompanying them, Fuseli's near-naked god envelops a bare-breasted woman within his luxurious wings.

In *The Loves of the Plants*, Darwin had explained his belief that a poet should be able to summon up scenes for readers to visualize, and *The Temple of Nature* is structured as a series of images. Like many of his contemporaries, Darwin was fascinated by Egyptian culture and language. In 1799, Napoleon's troops discovered the Rosetta stone, with its three versions of the same priestly decree—one in ancient Greek, one in Egyptian script, and one in hieroglyphs. The hieroglyphic code was not deciphered until after Darwin's death, but he was convinced that its signs concealed ancient chemical and historical knowledge. He wanted to create a universal language, a sort of Esperanto of the arts that would be universally understood like musical notation, alphabetic letters, or the ziz-zag trace of an automatic recording instrument. In a sense, Darwin's poem resembles verbal hieroglyphs—allegorical scenes strung together that affect the reader through their visual, emotive impact as well as their rational meaning.[54]

When you open *The Temple of Nature*, the first thing you come across is Fuseli's startling frontispiece. A woman displaying the aristocratic profile of a Greek statue draws back a veil to reveal an extraordinary female figure with three breasts. After scouring the text, I realized that Fuseli must have been inspired by this stanza (in which 'purfled' means decorated with an ornamental border):

> SHRIN'D in the midst majestic NATURE stands,
> Extends o'er earth and sea her hundred hands;
> Tower upon tower her beamy forehead crests,
> And births unnumber'd milk her hundred breasts;
> Drawn round her brows a lucid veil depends,
> O'er her fine waist the purfled woof descends;
> Her stately limbs the gather'd folds surround,
> And spread their golden selvage on the ground.[55]

If modern books carry any opening picture at all, it is usually a photograph of the author with little direct relevance to the contents. However, in Darwin's time, elaborately drawn frontispieces represented and summarized the writer's major theme by using allegorical imagery, as if using a symbolic visual language that could be decoded. For me, coming to Darwin and Fuseli two centuries later, both the verse and the picture were alien. Why should Darwin's Nature have a hundred breasts and arms?

Glancing down to the foot of the page, I saw a footnote about the Eleusinian mysteries, which Darwin clearly assumed that his readers would know about. Would exploring the Eleusinian mysteries resolve the more immediate mystery of this multi-breasted Nature? Initially I was unsure, but they proved to be extremely significant for helping me to understand Darwin's poetry and his attitudes towards science.

Mysteries of Science

The Eleusinian mysteries were unknown to me, but when I consulted my poet, he was surprised (well, horrified really) at my ignorance. So if you're a more knowledgeable reader than I was, please skip the next few paragraphs. For those of you who share my lack of a classical education, I hope that this brief summary of what I discovered will help.

I started by going back to my translation of Ovid's *Metamorphoses* and reading some more Greek myths. I had already encountered Persephone in *The Loves of the Plants*, and now I met her in a different context. Half-remembering the tale from lessons at school, I initially adopted the child-like approach of considering it as an unusually complicated fairy-story...

This beautiful young maiden had important parents—Demeter, an Earth Mother figure associated with agriculture and fertility, and Zeus, ruler of the gods. She also had a wicked uncle, Zeus's brother Hades, god of the underworld. One day, when she was tranquilly gathering armfuls of flowers in a meadow, a gaping crevasse opened up at Persephone's feet, and Hades pulled her down beneath the surface of the earth. Wracked with grief, Demeter searched desperately for her missing daughter, and ended up several adventures later disguised as an old woman in Eleusis, a few miles west of Athens. The king of Eleusis built a temple for Demeter, where she hid herself away and used her divine powers to spread a great famine over the land. Although ordered by Zeus to return his niece, Hades deceitfully tempted her to eat some pomegranate seeds, with the consequence (and here I omit several explanatory steps) that she could only return to her mother for part of the year before being recalled to the underworld and her wicked uncle. Overjoyed at her success, Demeter set off for Mount Olympus and Zeus, but before leaving Eleusis, she revealed her secret rites to a select group.

As a child, I had failed to appreciate (or perhaps hadn't been encouraged to think about) the symbolic importance of myths and their role in contemplating the most fundamental questions about existence. In this poem, Darwin uses various perspectives to address from different angles some key themes—life and death, good and evil, reproduction and sexuality, change and progress. For him, science and myth are complementary, not antagonistic, approaches for describing a pantheistic cosmos, one in which sexual energy fuels living creation as well as personal appetites.

Persephone's annual descent into darkness corresponds to the onset of winter, and her reappearance marks the arrival of spring. Originally the goddess of death, she is also resurrected to become the source of the life that emerges out of darkness. A few thousand years ago—and this is history, not mythology—a sacrificial cult dedicated to Demeter started to develop at Eleusis, which became one of the most important religious centres in ancient Greece. Packed with magnificent temples, Eleusis attracted pilgrims from far afield before Christianity became prominent and its adherents condemned as heretical this older religion of resurrection.

Extraordinarily, although special rites were performed annually for many hundreds of years, the content of these so-called Mysteries remained a close secret. It seems clear that there were fasts, feasts, and sacrifices, accompanied by carefully choreographed ceremonies of initiation conducted by a priest. Very probably, they involved sexual acts of initiation, but the details are obscure even today.[56]

Centuries later, these pagan rituals became symbolically significant for Darwin's contemporaries and the Romantic poets who followed them. Two hundred years ago, in the period of widespread anxiety and unrest after the French Revolution, rumours of Rosicrucian sects, Eleusinian mysteries, and diabolic mesmerizers were being whispered all over Europe. Conspiracy theories are seductive—one reason why *The Da Vinci Code* has fascinated so many readers. Authors speculated feverishly that international cartels ruled the world, that Jews and infidels were plotting to destroy European civilization, that the freemasons gathered at the local coffee house were secretly inciting political ferment. Even the sober *Encycloaedia Britannica* carried a preface warning patriotic Britons about the insidious machinations of French academics.

In the preface to *The Temple of Nature*, Darwin reminds his readers that the Eleusinian Hierophant, the head priest, taught chosen initiates 'the philosophy of the works of Nature', an ambiguous phrase that was redolent with sexual overtones and alarmed Darwin's critics. Unlike atheism, the pantheistic philosophies that Darwin and the Romantic poets expounded in the early nineteenth century challenged Christianity because they offered an alternative religion whose resurrection story was rooted in physical nature. 'Philosophy' summoned up associations with the atheistic French *philosophes* whom Burke and his supporters blamed for the horrors of the Revolution, while to talk about Nature's 'works' implied some sort of active force, an innate energy that is intrinsic to matter and operates on it. Such a view was anathema to orthodox Christians, who believed that activity could only be infused into matter by God: the natural world cannot run itself, but requires divine intervention in order to generate life.

As I tried to learn more about the Eleusinian mysteries, I encountered some familiar names. Experts at the time included Thomas Taylor, who had satirized Wollstonecraft but was here, as a Platonist scholar, on more familiar intellectual territory. Still more relevant to my Darwinian quest was Knight, one of several antiquarians who stimulated public interest in the Eleusinian mysteries around the end of the eighteenth century. In his controversial *Discourse on the Worship of Priapus* (1786), Knight claimed to have worked out these secret rituals from surviving texts. According to him, the elite members of the Eleusinian priesthood were scientifically way ahead of their time, even realizing that the sun, not the Earth, lies at the centre of the solar system. However, they deliberately concealed their superior knowledge from outsiders.

This two-tier structure of privileged participants and ignorant non-initiates was adopted by secretive cults such as Freemasonry that flourished all over Europe. It was not so very different from the new model of professionalized science that was starting to develop. Resembling Eleusinian Hierophants or Masonic Grand Masters, men of science were promoting themselves as highly trained experts who performed ritualized experiments for the benefit of their uneducated audiences. In *The Temple of Nature*, Darwin explicitly refers to these two groups, the vulgar plebs who are to be excluded and the special few with access to specialized knowledge:

> Hence ye profane!—the warring winds exclude
> Unhallow'd throngs, that press with footstep rude;
> But court the Muse's train with milder skies,
> And call with softer voice the good and wise.[57]

Every time I reread Darwin's poem, I penetrated a little further. Much of it is voiced by Urania, another of Zeus's daughters, the Muse responsible for astronomy and poetry. A few pages in, Urania conducts a favoured band of initiates into the inner shrine of Darwin's imaginary Temple with its majestic sculpture of Nature, illustrated by Fuseli. The towers on Nature's head are borrowed from Cybele, worshipped as an Earth Mother, while her multiple breasts refer to the statue of Diana that used to stand at Ephesus, now in Turkey but then a major Greek trading post lying between Europe and Asia.

Diana was the name given by the Romans to the Greek goddess Artemis, another of Zeus's daughters and hence Persephone's cousin. I unearthed several academic articles about her multiple breasts, each offering their own esoteric interpretation—tassels, hills, testicles (Cybele was the Phrygian version of Aphrodite, whose priests castrated themselves). Whatever they were originally meant to be, by Darwin's time there was no uncertainty. Artemis represented both chastity and fecundity: rather like the Virgin Mary in Christianity, she was simultaneously unsullied goddess and fertile Nature. Conveniently for fountain-makers, water could spurt out from the breasts of Artemis, who became an important icon in Revolutionary Paris because women were being encouraged into domesticity, exhorted to nurture the children of France by dispensing with wet nurses to suckle their own babies.[58]

Although at first they had seemed strange to me, the frontispiece and its accompanying lines present a central message of Darwin's book. The Eleusinian mysteries stood for the truths of nature, decipherable only by the Hierophant's initiates—men like himself—who had the wit to understand them. For explaining his discoveries to less privileged people—his readers—he gave Nature a hundred breasts to represent two basic aspects of his theories: sustenance and reproduction. Following Lucretius' lead, Darwin intended the towers piled up on Cybele's forehead to indicate that she carries the weight of cities. This allegorical imagery corresponds to Darwin's alternative title, *The Origin of Society*, and conveys poetically his conviction that nature is the basis not only of life, but also of human civilization.

Seductively elusive, the Eleusinian mysteries were starting to dominate my life. To do research effectively, it helps to be the sort of obsessive who reads the footnotes before the rest of the book, and who delights in days spent pursuing fragile clues. But only too easily those days can stretch into weeks, and then into months or even years. It was time for some self-assessment. I knew I would never become an expert on Greek mythology, hieroglyphic symbolism, or Lucretian poetry. Darwin included such references because he was struggling to place his own ideas in familiar contexts. But for me, they were proving an obstacle, not an aid, to understanding what Darwin thought about evolution and why his poem seemed so threatening to his peers.

For the time being, I decided, I would focus on the shape of his poem and his basic intellectual messages. Once I had grasped those, I would explore their implications more fully.

A Very Quick Tour through Darwin's Temple of Nature

Recalling Ovid's four ages of the world, or the Enlightenment stages of human history, *The Temple of Nature* is divided into four sections. But Darwin was being more ambitious than even the most enlightened progressivist Scot—he wanted to sketch out not only the development of society, but also the emergence of life before civilization began.

Logically enough, Darwin's first canto is called 'Production of Life'. Straightaway, he plunges directly into a contentious topic that his grandson Charles would later cautiously and consistently avoid: the creation of life. According to conservative Anglicans who interpreted the Bible literally, God not only created the Earth but also established it exactly as it is now, complete with plants, animals, and human beings. In contrast, like many other Christians of his time, Darwin was a deist who gave God the more restricted role of setting the universe off at the beginning but then stepping back, so that all subsequent developments were governed by natural laws.

Flowery verse lends itself to vague generalizations, but Darwin boldly declares, without prevarication, that life has emerged from matter. For his contemporaries, his opening lines were scandalously heretical:

> By firm immutable immortal laws
> Impress'd on Nature by the GREAT FIRST CAUSE.
> Say, MUSE! How rose from elemental strife
> Organic forms, and kindled into life;[59]

Before Nature, there was Love. The Bible says that 'God is Love', but Darwin reverses this, so that Love is God. Echoing Lucretius and Milton, Darwin wrote:

> IMMORTAL LOVE! who ere the morn of Time,
> On wings outstretch'd, o'er Chaos hung sublime;
> Warm'd into life the bursting egg of Night,
> And gave young Nature to admiring Light![60]

What sort of LOVE did Darwin have in mind? By describing it as IMMORTAL, Darwin implies that it is in some way divine. On the other hand, Fuseli's illustrations and Darwin's ambiguous vocabulary suggest that for him, *eros* was at least as important as *agape*. It is sexual love that prompts living beings to create offspring and reproduce themselves. This unabashed celebration of sexuality antagonized his critics still further.

Darwin devotes his second canto to 'Reproduction of Life'. Perhaps thinking of his own ageing, he starts with death, explaining how deceased bodies are continually broken down and recycled so that new ones can be kindled into being. As ever, specificity is not his strong point, and the processes of generating successors remains unclear:

> But REPRODUCTION, when the perfect Elf
> Forms from fine glands another like itself,
> Gives the true character of life and sense,
> And parts the organic from the chemic Ens.[61]

Before long, Darwin has glossed over the details and skated on to a more familiar terrain of explicitly sexual imagery, progressing through creation from primitive sea-creatures up to flowers, insects, and animals. His Garden of Eden is an erotic playground, the site not of temptation but of jubilant fornication. Although there are some obvious candidates for anthropomorphization—heroic cockerels, regal kings—Darwin even manages to enthuse about the sex life of snails. This second canto is packed with romantic clichés—virgin bosoms, lovesick Beauties, rosy lips—and when Hymen presides over the marriage of Cupid and Psyche, he shouts out a prayer to sexual enjoyment:

> 'Behold, he cries, Earth! Ocean! Air above,
> And hail the DEITIES OF SEXUAL LOVE!
> All forms of Life shall this fond Pair delight,
> And sex to sex the willing world unite.'[62]

After all that exuberance comes the third canto, 'Progress of the Mind', which celebrates the rise of reason and the triumph of language. Even here, despite the shift towards affairs of the brain, Darwin contrives to squeeze in several titillating images of maidenly ivory shoulders and sleep-blushed cheeks. He also renders into verse a theory he had developed in *Zoonomia* that explained (well, not really: implied, hinted or indicated would be more accurate) how physical actions can result in mental activity. Gaining deceptive authority from some scientific-sounding words—stimulations, fibrous motions, contractions—Darwin traces evolutionary development, describing how creatures can affect their environment by building nests or trapping prey. Including 'the reasoning reptile' among our relatives, Darwin declares provocatively:

> —Stoop, selfish Pride! survey thy kindred forms,
> Thy brother Emmets [Ants], and thy sister Worms![63]

According to Darwin, human beings are different from other animals because they can communicate and plan ahead:

> To express his wishes and his wants design'd
> Language, the *means*, distinguishes Mankind;[64]

For his last canto, Darwin takes a different tack. Called 'Of Good and Evil', it celebrates human achievements (unsurprisingly, mainly those of his Lunar friends), but it also sketches out a grim picture of constant strife, a battle for survival in which every animal's existence depends on the death of others. Sounding like a poetic precursor of his grandson, Darwin writes angrily:

> From Hunger's arm the shafts of Death are hurl'd,
> And one great Slaughter-House the warring world![65]

In the poem's grand finale, Urania draws aside Nature's veil, sinks to her knees and lifts 'her ecstatic eyes to TRUTH DIVINE!' But any assiduous readers who persevere this far will discover that they are nowhere near the last page. Instead, there follow substantial chunks of prose, which cover topics as diverse as electricity, reproduction, taste, and volcanoes (as well as 'storge', which I found out means family love). If you add in the page-by-page footnotes, Darwin's technical commentaries take up over half his book. Although mixing genres in this way was more common then than now, *The Temple of Nature* is an extreme example of presenting unorthodox ideas within a double wrapping of mythological verse and scholarly hypothesis.

The concluding pages are perhaps the most astonishing. Here Darwin considers the 'Analysis of Articulate Sounds', explaining how each language involves slightly different combinations of lips, throat, and mouth positions. Inspired by some meticulous research, Darwin constructed an artificial talking head made from wood, leather, and silk ribbon. Like many other enthusiastic inventors, Darwin never quite got round to specifying the final details, but he did at least persuade his speaking-machine to produce simple syllables. Perhaps he would eventually have achieved his ambitious plan of designing a mechanical loudhailer powerful enough to control a crowd. He might even have fulfilled the joke contract he signed with Matthew Boulton for 'an Instrument called an organ that is capable of pronouncing the Lord's Prayer, the Creed and Ten Commandments'.[66]

Reluctantly putting thoughts of his speaking head to one side, I realized that to learn more about Darwin's evolutionary ideas, I needed to understand his contemporaries' answers to that basic and still unresolved question: How did life begin?

CHAPTER TWELVE

Origins

But catastrophes only encouraged experiment.
As a rule, it was the fittest who perished, the mis-fits,
forced by failure to emigrate to unsettled niches, who
altered their structure and prospered.

W. H. Auden, 'Unpredictable but Providential
(for Loren Eiseley)' (1976)

Young ladies were enticed into Darwin's imaginary garden, but they were not supposed to write books of their own. Mary Shelley's *Frankenstein* appeared anonymously, and even decades later the three Brontë sisters thought it wise to adopt male pseudonyms. In 1797, while Darwin was working on *The Temple of Nature*, an unknown aspiring authoress naively submitted *First Impressions* under her own name. When the publisher promptly rejected her manuscript without even reading it, he missed a great opportunity—this was an early version of *Pride and Prejudice* (1813), the novel with the most famous opening in English literature. 'It is a truth universally acknowledged,' wrote Jane Austen, 'that a single man in possession of a good fortune, must be in want of a wife.'

In 1789 the French chemist Antoine Lavoisier had used the same formula as Austen, the same rhetorical flourish, to introduce his manifesto for a revolutionary chemistry (he was later guillotined). 'It is a maxim universally admitted in geometry,' wrote Lavoisier, 'that...'. By appearing to state the obvious, Lavoisier left little opportunity for disagreement, and in its own first sentence *Pride and Prejudice* mocks the arrogance of such sweeping scientific statements.[67] Since the mid-seventeenth century, when the Royal Society was founded, natural philosophers had claimed to distinguish themselves from other writers by using non-metaphoric language and stating only what they knew. In practice, their arguments incorporated strategies of classical rhetoric, and their writing bore more resemblances to fictional genres than they were willing to acknowledge. While Darwin was agonizing over whether he should enlist imagination under the banner of science, scientific authors were relying on persuasive language to present their case, and novelists were attempting to describe real life.

How safely can historians extrapolate from the imagined worlds of novels to the actual experiences of historical actors? This question has generated heated debates amongst academics. Hard facts are clearly crucial, but Austen's characters do feel endearingly—or infuriatingly?—realistic. When you immerse yourself in her stories, you step back two centuries into tight-knit communities fraught with emotional intensity and conflict. They were probably not so very different from the gossipy circles of Lichfield, which—like other provincial cities such as Nottingham and Shrewsbury—had a high proportion of widows and spinsters.[68] Local diaries reveal that in the closely packed surroundings of the city's cathedral close, animosities reminiscent of a soap opera simmered for years. Lichfield's doctor was not above joining in.

Not the most tactful of men, Darwin accumulated several enemies, but a particularly vituperative neighbourly dispute erupted in 1770, when he decided to modify the family coat of arms on his carriage. To supplement its three scallops (symbols of fertility and pilgrimage), Darwin added three words—*E conchis omnia*, Latin for 'Everything from shells'. Retaliation was swift. Canon Seward, Darwin's one-time benefactor and Anna's father, took great exception to this slogan. Resorting to poison-pen poetry, he accused Darwin of abandoning Christianity and instead following Epicurus, Lucretius' hero, the Greek philosopher who believed that the cosmos is ruled not by God, but by chance:

> He too [i.e. Darwin] renounces his Creator,
> And forms all sense from senseless matter.
> Great wizard he! by magic spells
> Can all things rise from cockle shells...
> O doctor, change thy foolish motto,
> Or keep it for some lady's grotto.[69]

All pretence of friendship between the two men evaporated. Darwin did resentfully remove the offending phrase from the side of his carriage, but he kept it on his bookplate. I was mystified. Shells were a collectors' item in the eighteenth century, patiently accumulated to decorate grottoes or purchased (sometimes at vast prices) for arranging tastefully in cabinets. Darwin was friendly with Richard Greene, the local apothecary, whose Museum of Curiosities attracted eminent visitors such as Johnson and Boswell to admire its display of shells.

But that did not explain why Darwin and Seward both attached so much significance to this short slogan. My immediate instinct was to bury myself in the Rare Books Room at the library and find out. But life intervened, as it does...and I put this research on hold.

Creation

Fortunately, serendipity intervened as well, and during my enforced separation from Darwin, I stumbled across some clues. The first was the frontispiece of a 1651 book about reproduction by William Harvey, the seventeenth-century anatomist. Above the title, Zeus sits on an eagle, holding the two halves of an egg labelled *Ex ovo omnia*, or 'Everything from the egg'. That sounded familiar. Presumably, as a doctor Darwin had Harvey's pronouncement in mind when he composed his own dictum, *E conchis omnia*, or 'Everything from shells'; his well-educated colleagues would immediately have picked up the reference.

Harvey is often said to be 'The Founding Father of Modern Medicine', because he proved that blood is not continuously generated in the liver, but instead circulates around the body (although this discovery would have little impact on medical practice for another couple of hundred years). More relevantly for me, Harvey was also fascinated by reproduction. To investigate the mechanisms of conception, he took advantage of his privileged position as the king's physician by observing the deer in the palace park. On the basis of his research, Harvey set out a seventeenth-century version of what is now a fundamental principle of animal biology: that offspring only start to be formed when male semen interacts with a female egg.

In retrospect, Harvey seems to have got the right answer—but significantly, not everybody believed him at the time. Even a couple of hundred years later, in the nineteenth century, some people were ready to believe that life could be generated spontaneously. A national scandal erupted after the report that Andrew Crosse, a wealthy Somerset landowner, had created life with electricity. He had been trying to produce crystals by electrifying a small stone brought back from Mount Vesuvius, and was astonished one day to see tiny insects wriggling around inside his apparatus. Deciding that was a more interesting research topic, Crosse started exploring the conditions needed to produce these living mites, which—according to his own drawings—had segmented legs and smooth backs covered with long curvy bristles.

Unfortunately for Crosse's tranquillity and reputation, a West Country journalist decided to boost sales by writing a dramatic but not entirely accurate article that got picked up by the London press. By January 1837, Crosse had become an early victim of media sensationalism. Fans hailed him as a genius, the hero of modern science who had used electricity to show once and for all that life originated from combinations of atomic building blocks. In contrast, sceptics accused him of sacrilegiously taking over God's unique ability to produce living beings. 'Ridicule is the only weapon we can condescend to use,' declared Cambridge University's professor of geology, 'against...this mockery of a creative power.' This orthodox lobby proved more powerful than Crosse, and his claims were dismissed, but the very virulence of the controversy demonstrates that uncertainties did still exist.[70]

Between Harvey and Crosse—*via* Darwin and many others—lay two centuries of Europe-wide debates about reproduction. The full story is tortuous, but basically opinions polarized into two main opposing positions: preformation and epigenesis. According to the preformationists, every creature currently alive stems from an original act of creation by God. The first member of each species contained within it all future offspring, so that like Russian dolls being unpacked, latent embryos exist already fully formed: effectively, miniature babies simply get bigger before being born. Subsequent generations may be very different from their ancestors living today, but any changes have been planned in advance by God.

In contrast, Harvey was an epigenesist, because he believed that at conception, some sort of power in male semen acts on an egg to generate an embryo and encourage it to develop. As critics pointed out, proponents were rather vague about what that power might be. Buffon, for instance, wrote much about 'penetrating forces' and 'internal moulds', but it was hard to make those nebulous yet purposeful concepts compatible with inanimate Newtonian forces. Darwin argued that the mother provides nourishment for an embryo supplied by the father, but could provide no theoretical explanation.[71] Yet the preformationists' own theories also lacked precise detail and proved hard to confirm experimentally.

The debate between these two camps was as much about theology as biology. Because epigenesists envisaged successive acts of reproduction without God's direct involvement, it was easy for preformationists to accuse them of favouring atheism and materialism. When Darwin was writing his *Temple of Nature*, evidence was piling up to support epigenesis. The religious implications of this research deterred potential experimenters, but even so, laboratory work confirmed that the tissues and organs in a foetus appear gradually, becoming progressively more specialized and more distinct—or to use Darwin's own term, they evolve. As he set about attacking the preformationist opposition, Darwin referred dismissively to those 'ingenious philosophers' who suppose 'all the numerous progeny to have existed in miniature in the animal originally created'.[72]

Yet despite all the experimental research, it remained obscure how patterns of growth are caused and controlled. Why does one fertilized egg grow into a fish and another into a horse? How does a tiny egg differentiate itself into legs, kidneys, and a brain?

One extraordinary incident illustrates the miasma of mystery surrounding both conception and pregnancy. In 1726 even normally sedate newspapers carried reports that a country villager called Mary Toft was repeatedly giving birth to baby rabbits. Some of London's most eminent physicians travelled down to Surrey and examined her for themselves—and some of them vouched that her story was true. Gullible as these doctors might seem, their behaviour was less laughable than it would be now.

According to one respected medical theory, a mother's mental state affects her embryo's development. Wealthy women were routinely encouraged to soothe their forthcoming babies by relaxing, reading pious literature, or listening to soft music, while those poorer mothers obliged to keep working were warned about the dangers of seeing a hare, which could give rise to a cleft palate, popularly known as a harelip. Although Darwin denied the effects of a mother's imagination on her foetus, he did attribute to fathers the power of affecting the sex and skin colour of their children.[73] Toft explained that she had already miscarried a human baby after being startled by a rabbit running across a field—and ever since, she had been obsessed by the thought of rabbits. This account sounded credible to advocates of maternal influence during pregnancy.

Eventually Toft confessed to having started a hoax that had spun out of control, but uncertainty about the power of a mother's experiences survived well into the nineteenth century, as is made clear by this Scottish joke about cuckolding. 'I have known a lady', remarks a naive gentleman, 'who was delivered of a blackamoor child, merely from the circumstance of having got a start [shock] by the sudden entrance of her negro servant, and not being able to forget him for several hours.' His servant was savvier: 'It may be, sir; but I ken this;—an I had been the laird, I wadna hae ta'en that story in.'[74]

Approaching the puzzle of development from a different angle, some eighteenth-century researchers investigated what happens when you operate on living animals. Unconstrained by modern ethical regulations, they found that lobsters regrow their claws, that salamanders produce new tails, and that two polyps (tentacled water creatures) are produced by cutting an original one in half. Such results suggested that animals possess some sort of inner vital force, as though material flesh is imbued with a spiritual ability to regenerate itself. For devout Christians—especially Calvinists committed to divine predestination—this notion verged dangerously close to materialism, atheism, and the displacement of God from the universe.

Spontaneous generation seemed still more threatening. Microscopes revealed that even the cleanest of liquids contain myriad swarming animalcules, but nobody was sure what they were or how they originated. Did they transmit diseases? When food went rotten, which came first— did insects cause meat and fruit to decay, or was it the decay that produced the insects? People tried to find the answers by boiling up broths and watching what happened during the next few weeks, but the results were inconclusive. Some experimenters maintained that tiny creatures arrived as if from nowhere, but their opponents had an easy riposte: they must have flown in from the surrounding air through leaks in the apparatus.

Could creation be a natural event taking place without God's intervention? Could the living world be explained by a materialist creed? Despite censorship, these suggestions circulated in France from the middle of the eighteenth century. As well as running counter to religious beliefs, they also carried great political significance. The hierarchical structure of European societies reflected God's foundation of a stable system governed from the top down. If nature was in charge, and matter could generate life without divine intervention, then traditional arguments would no longer be valid. The entire edifice might crumble—as indeed it had, during the French Revolution. In Britain, anxious to prevent radicalism from seeping over the Channel, Darwin's contemporaries emphasized the subversiveness of supporting spontaneous generation.[75]

Darwin's own reservations about the possibility stemmed from his faith not in God, but in enlightened rationality. If living creatures can suddenly appear, apparently at random, then the universe is not governed by regular laws, and so is not subject to the power of reason. In that case, who is to say that cows might not start growing six heads? Not prepared to abandon fixed rules of nature, Darwin flatly rejected the idea that large animals could be spontaneously generated in their entirety. He did, however, accept Buffon's suggestion that a simple form of life originated spontaneously from a primeval chemical soup. Darwin proposed that there had been one single initial moment, many many aeons ago, when God created life from matter. Writing tentatively, in *Zoonomia* he wondered whether 'in the great length of time, since the earth began to exist, perhaps millions of ages before the commencement of the history of mankind, would it be too bold to imagine, that...animals have arisen from one living filament, which THE GREAT FIRST CAUSE endued with animality...'[76]

Because of its political and religious ramifications, spontaneous generation was an emotive term. Critics found it easy to condemn *any* suggestion that life could appear by itself, but there were two distinct positions. The more extreme of these, denied by Darwin, was to claim that materialist processes of creation have been happening continually ever since the beginning of the world. Instead, Darwin and many German researchers adopted the less scandalous view that there was just one single moment when an initial living germ was generated: all other creatures have descended from it. Although a substantial group of naturalists held this opinion, they were later written out of history by Charles Darwin and his allies. This deliberate erasure made evolution by natural selection look much more attractive, because it was presented as the only viable alternative to repeated miraculous creation.[77]

In *The Temple of Nature*, although God initially creates the cosmos, Nature immediately takes over so that life emerges as a consequence of physical causes. First the planets are formed by being hurled out from flaming Chaos, and then:

> Nurs'd by warm sun-beams in primeval caves
> Organic Life began beneath the waves.[78]

In the next few pages, by supplementing his verse with scholarly footnotes, Darwin conveys the impression that he is providing a full explanation. After much rereading, I concluded that he hides behind poetic imagery and never commits himself to a precise mechanism. One deceptive tactic he uses is repetition—only sixty lines later, I found the same sentiment, the same rhymes:

> ORGANIC LIFE beneath the shoreless waves
> Was born and nurs'd in Ocean's pearly caves...[79]

Persevering through the couplets, I deduced that his filaments are infused with life by heat, attraction, and contraction so that (somehow) they become woven into the fabric of nature. Rather than berating Darwin for not coming up with a fully fledged theory of creation, I decided it was more interesting to reflect on his courage in going public with such an unorthodox position. Amongst his friends, he had long been open about his controversial views. The daughter of his Lunar friend Samuel Galton remembered that 'Dr Darwin often used to say...Man is an eating animal, a drinking animal, and a sleeping [euphemism?] animal, and one placed in a material world, which alone furnishes all the human animal can desire. He is gifted besides with knowing faculties, practically to explore and to apply the resources of this world to his use. These are realities. All else is nothing; conscience and sentiment are mere figments of the imagination.'[80] Making such a declaration at a domestic dinner party is very different from committing it to print—but that is what Darwin did.

Perhaps fortunately for his pride, Darwin died before *The Temple of Nature* was published, but he may sometimes have thought back to his row with Seward. Redesigning the coat of arms on his carriage might seem a trivial act of rebellion, but Darwin's provocative slogan *E conchis omnia*— 'Everything from shells'—suggests that he had already redesigned his ideas about the origins of life forty years earlier.

Evolution

In his more optimistic moments, Darwin may have flattered himself that his ideas would revolutionize biological thought and ensure him a place in the scientific pantheon. He had no idea that, thanks to his grandson, his surname would be forever indelibly linked with evolution. But what exactly were Erasmus Darwin's own views? Deciding to make yet another attempt at deciphering *The Temple of Nature*, I ordered up one of the four original copies held in Cambridge University Library. Inadvertently, I picked the one owned by Charles Darwin. 'Ch Darwin' it said on the flyleaf—and there were his handwritten notes! I started skimming through, searching for any clues to his impressions of the grandfather he had never met.

I soon came across several pencilled marks in the margin, which (I checked later) are also by Charles Darwin.[81] He had evidently paused over this particular suggestion by his grandfather: 'Perhaps all the productions of nature are in their progress to greater perfection!'—a surmise I recognized as recycled from a long footnote on turmeric in *Loves of the Plants*.[82] At the very beginning of his literary career, Erasmus Darwin was already fretting over apparent redundancies such as male nipples, wondering whether they could be vestigial traces left over from long processes of transformation stretching back for aeons. By the time he came to write

The Temple of Nature, he had been sifting these ideas carefully, keeping up-to-date with the latest theories of continental scholars such as Blumenbach and Buffon, repeatedly modifying his arguments to accommodate the latest observations and suggestions.

By the end of the eighteenth century, the notion of change was no longer in itself especially scandalous. For several decades, the word 'evolution' had been in use for living beings, and there were several strands of evidence arguing against a literal interpretation of the Bible. Giant fossils—such as mammoths and giant elks—suggested that the world had once been inhabited by distant relatives, now extinct, of familiar creatures. Animal breeders reinforced particular traits to induce changes carried down through the generations—stalwart bulldogs, athletic greyhounds, ladies' lapdogs. Geological data was also accumulating: seashells on mountain peaks, earthquakes, strata lacking fossil remains—and the most sensible resolution for such puzzles was to stretch out the age of the Earth and assume that it is constantly altering.

Many people accepted that the present living world is very different from the one originally created by God. Linnaeus, a pastor's son, ingeniously reconciled Scripture and observation by proposing that the diversity of plants around the globe had not always existed. Instead, he argued, God initially created only a few plants, but over generations, multiple floral marriages (his terminology) caused the number of varieties to increase. What horrified Christians was the suggestion that there was no divine plan governing the pattern of change. Charles Darwin's inspiration was not to invent evolution, but to explain how it could work. Seventy years before he dared publish *On the Origin of Species*, his own grandfather was already rejecting permanence and backing transformation. Even he was not the first to do so.

Returning to *The Temple of Nature*, I focused on the first two cantos, which discuss respectively the production and reproduction of life. Darwin's vision of evolution was buried beneath poetic evocations of tawny lions ('bride' conveniently rhymes with 'pride'), copulating snails overwhelmed by their desires ('perplex' and 'sex'), and bashful dryads peeping out at emerging nereids and enamor'd tritons ('coral cells' and 'twisted shells'). The following lines provide a relatively coherent summary of his vague plan that life forms have developed over time, starting with invisible marine creatures and then progressing upwards through plants to fish and mammals:

> First forms minute, unseen by spheric glass,
> Move on the mud, or pierce the watery mass;
> These, as successive generations bloom,
> New powers acquire, and larger limbs assume;
> Whence countless groups of vegetation spring,
> And breathing realms of fin, and feet, and wing.[83]

This evolutionary sentiment had provoked outrage in *Zoonomia*, where Darwin had speculated that 'all warm-blooded animals have arisen from one living filament... continuing to improve by its own inherent activity, and... delivering down those improvements by generation to its posterity, world without end!'[84] The *Anti-Jacobin* tore his filaments into shreds, imagining the universe as an original point that extends into a line until vegetables grow wings, algae become fish, and people rub off their tails. And over sixty years later, it was this passage from 'The Loves of the Triangles' that Bishop Wilberforce chose to quote in his attack on Charles Darwin's idea of natural selection.[85]

In *The Temple of Nature*, Darwin reiterates the ambitious Theory of Just About Everything that he first presented in *Zoonomia*. There he drew a parallel between two powers: gravitational attraction, which governs the inanimate world, and contraction, controller of living creatures. Typically for aspiring philosophers of this period, Darwin set out his definitions and axioms with apparently mathematical precision. Even so, clarity is not one of his model's strong features. Schooled in a Newtonian approach, Darwin sought to establish causal, mechanical links between the physical and the mental, between bodily sensations and psychological emotions. His technical explanation was built up on a fourfold foundation of Irritation, Sensation, Volition and Association—terms that sound unfamiliar now, but that cropped up frequently in medical and biological texts of the period.

Darwin derived his first two fundamental capacities directly from the Swiss anatomist and physiologist, Albrecht Haller (also a doctor who wrote poems about nature): the irritation that arises from contact with things, and is crucial for processes such as digestion and circulation, and the sensation resulting from pleasure or pain. For the other two, he also borrowed from the philosopher David Hartley, who tried to explain how mental desires and physical actions could be linked. According to his doctrine of associations, which was extremely influential, vibrations in the brain and nerves are affected by past activities as well as present experiences, so that ideas and sensations become intimately linked by association with one another. Darwin identified volition, the desire to experience or avoid something (although this may be unconscious, as when you unthinkingly brush away an annoying fly), and association, which occurs when one action automatically triggers off another. His examples include learning how to dance, animals' in-built responses to outside threats, and the aesthetic appreciation of curves stemming from breastfeeding.[86]

By combining his observations and his theories, Darwin confirmed his belief that all living beings contain within themselves the capacity to respond and to improve. Often unwittingly, we are constantly monitoring our interaction with the environment: when we squeeze a coin, after some time we no longer notice it even though the physical contact remains. According to him, Asians are slighter than Europeans and Africans because they are neither agricultural labourers nor proficient swimmers: broad-shouldered men produce broad-shouldered children. Mind and body interact, as Darwin knew from his personal experience of losing his stutter when on his own.

In *The Temple of Nature*, Darwin imagines how his four faculties actively engaged in fashioning animals' characteristics as they gradually emerged over the millennia to culminate in the human race. Charles diplomatically omitted people from his evolutionary scheme, but Erasmus was bolder: he makes it clear that we have evolved naturally from animals rather than being created separately by God, a view consistent with his abolitionist politics. We too, he insists, have emerged from microscopic origins (as in *Zoonomia*, he uses the word 'ens', an unusual and not particularly informative term denoting something that exists in its own right):

> Imperious man, who rules the bestial crowd...
> Arose from rudiments of form and sense,
> An embryon point, or microscopic ens![87]

Instead of skimming over this contentious topic, in the third canto Darwin emphasizes still further the continuity of animals and humans. In one place, he likens an angry lion lashing its tail to a 'Savage-Man' parading around with clenched fists; in another, he emphasizes human deficiencies— we can neither run as fast as a fox nor soar as high as a hawk, nor smell as acutely as a hound. In short, he makes our animal origins explicit.

As well as exploring our shared physical ancestry, Darwin examined mental resemblances in order to remove the conventional barrier of reason separating people from animals. The *Anti-Jacobin* poets would presumably not have approved of these couplets describing how our volitions (roughly equivalent to decision-making abilities) relate us to lowly forms of life:

> Wise to the present, nor to future blind,
> They link the reasoning reptile to mankind!
> —Stoop, selfish Pride! survey thy kindred forms,
> Thy brother Emmets [Ants], and thy sister Worms![88]

As his grandson would do in the following century, Darwin examines the animal world closely, seeking insights into human behaviour by describing how elephants pick up coins, squirrels run around their cages, and birds line their nests with moths. He was particularly struck by watching a wasp, which deliberately tore the wings off a fly so that it would be easier to carry (Charles Darwin later borrowed that observation for his own notebook). Puppies and babies alike, he remarks, investigate objects with their mouths (predictably, breastfeeding gives him the opportunity for an appreciative digression into lines of beauty).

Most importantly for Darwin, as the plan of creation rolled out towards perfection, two features arose that distinguish humans from all other creatures: our hands, with their sensitive fingertips and opposed thumbs, and our verbal language of communication. Thanks to the human sense of touch, we can both explore and alter our environment more effectively than can other animals. And because of our linguistic skills, we have come to dominate the world through our reason—we can remember, we can reflect, and we can plan for the *future* (one of the very few places in this poem where Darwin uses italics for emphasis).

The climax of Darwin's progressive poetic evolution occurs when a male seraph called Sympathy descends to civilize 'savage man' by imbuing him with morality. In this final stage of human development, people bond together in societies and recognize the virtues of giving to the poor or of liberating prisoners and slaves. Significantly, Darwin nowhere distinguishes between types or nationalities of human beings. For him, the chained African on Wedgwood's plaque is innately as superior to animals as he is himself. Translated into modern terms, Darwin interpreted the motto 'Am I not a man and a brother?' (see Fig. 7, p. 180) as 'Surely I too belong to *homo sapiens?*'

Anti-Jacobins rejected evolutionary ideas because they threatened two traditional certainties: Britain's hierarchical political system, with its inbuilt assumptions of hereditary superiority; and a stable cosmos created by God. Although Darwin's own work was panned (*Zoonomia* even found its way onto the Vatican's proscribed list of books), evolution seemed sufficiently feasible to remain controversial. As an elderly man, the naturalist Joseph Banks, president of the Royal Society, gave it serious consideration. In a letter to Richard Payne Knight's brother, a fellow botanist, Banks launched into a meandering chicken-and-egg-type witticism that turned on the conundrum of how lowly lice could have survived while waiting for their human hosts to evolve. Suddenly he adopted a more sober tone: 'untill [*sic*] an actual experiment has taught us that an animal can proceed from another without having been created or begotten, what inducement can we have for believing that possible from abstract reasoning, which appears impossible from actual experiment?'[89]

That very same criticism would later be levelled against Charles Darwin, who—like his grandfather—argued by accumulating evidence rather than providing a hypothesis that could be tested experimentally.

Improvement

Improvement inevitably involves change, which many people find hard to contemplate. 'Change is not made without inconvenience,' Samuel Johnson reminded readers of his *Dictionary*, 'even from worse to better.'[90] Such resistance contributed to the lack of revolutionary activity in Britain: for the lower classes, remaining in servitude seemed preferable to disrupting life as they knew it. If that seems unlikely, then reflect on the longevity of the Soviet regime. After it collapsed, many middle-aged Russians hankered after the security they had always known. Initiative, ambition, and self-sufficiency were attitudes they had never acquired, and even Joseph Stalin is now being remembered nostalgically by some.

In contrast, Enlightenment reformers believed in progress and dreamt of a better future. There was little distinction between science and politics, ethics and literature. Darwin's friend Thomas Day, the most prominent political campaigner of the Lunar Society, preached that mental improvement—education—was the key to removing oppression: 'he therefore that would save, must first enlighten'.[91] Six months after the Bastille fell, Darwin expressed his own hopes by enthusing to Watt: 'Do you not congratulate your grand-children on the dawn of universal liberty? I feel myself becoming all french [sic] both in chemistry and politics.'[92] Whether it involved manufacturing gases to launch balloons, or cultivating sugar beet to reduce the demand for American imports, he had faith that life would improve.

If Darwin's had been a lone voice, then he would not have aroused such widespread hostility: there is little point in sustaining prolonged attacks on an insignificant eccentric. Darwin's flamboyant verses, best-seller status and social prestige made him an appealing target for ridicule, but he represented a substantial swathe of British opinion towards the end of the eighteenth century. In particular, Darwin's declared ideology of improvement paralleled those voiced by William Jones and Richard Payne Knight.

Writing to Banks from India, Jones suggested how a species of Bengalese tree could be improved by cross-breeding, and wondered whether pangolins might represent some intermediate evolutionary stage towards reptiles.[93] In his poetry, Jones gave his British readers a modernizing incentive for tackling Hindu theology: Hinduism, wrote Jones, was 'devoutly believed by many millions, whose industry adds to the revenue of *Britain*'.[94] Like Darwin, he regarded creation as 'rather an *energy*, than a work',[95] and presented a cyclical model of natural growth and decay:

> Dense earth in springing herbage lives,
> Thence life and nature gives
> To sentient forms, that sink again to clay.[96]

A committed Christian, Jones steered away from suggesting a new model for the universe. Although his influential works on linguistics escaped the harsh criticism meted out to Darwin and Knight, his model of philological change paralleled theories of biological evolution: it contradicted the account of stability given in the Bible, and it undermined European assumptions of superiority. According to the book of Genesis, God created the present diversity of languages only when human beings had the temerity to build the giant Tower of Babel. But in Jones's view, older languages were superior to the classical Greek and Roman that emerged later. This focus on degeneration makes Jones the linguistic equivalent of Buffon, who suggested that European species deteriorated when they were stranded in the Americas.

Jones derived his methodology from botany. Using the Linnaean concept of families, he classified and described groups of languages to set up an evolutionary model of change and development. On the basis of his systematic quantitative studies, Jones concluded that the classical languages of India and Europe are descended from a common source. In his most famous passage, Jones wrote:

> The *Sanscrit* language ... [is] more perfect than the *Greek*; more copious than the *Latin*, and more exquisitely refined than either, yet bearing to both of them a stronger affinity ... than could possibly have been produced by accident; so strong indeed, that no philologer could examine them all three, without believing them to have sprung from some common source, which, perhaps, no longer exists ... the old *Persian* might be added to the same family.[97]

The third poet of the Lucretian triangle, Knight, endorsed progress through invention. Here he describes how humans acquired power through creating tools:

> Still one invention to another leads,
> And art to art, in order slow, succeeds: ...

Thus more effective implements were found
To raise the building, and to till the ground;
Labour by art was methodized and fed;
And man's dominion over nature spread.[98]

This vision of technological improvement was parodied by the *Anti-Jacobin*, which imagined a 'rude savage' making a sudden evolutionary leap from vegetarian forager to calculating hunter on seeing a tiger devour its prey (the reference to broad rolling eyes was intended to suggest an African):

Struck with the sight the wondering Savage stands,
Rolls his broad eyes, and clasps his lifted hands...
From the tough yew a slender branch he tears,
With self-taught skill the twisted grass prepares...
Twangs the bent bow—resounds the fateful dart,
Swift-wing'd, and trembles in a porker's heart.[99]

What made Darwin especially vulnerable to attack was his readiness to continue writing provocatively. In 1789, the year of the French Revolution, the astronomer William Herschel (discoverer of Uranus) published an innovative paper rejecting the entrenched notion of cosmic stability. From his observations of nebulae (cloud-like clusters of stars), Herschel concluded that ours is not the only galaxy, that the universe has been constantly changing, and that some galaxies are far older than others. In Herschel's luxuriant celestial garden (his imagery, not mine), the stellar equivalents of centuries-old oak trees and short-lived shrubs exist together. Still more frighteningly, the entire cosmos may flourish and decay, like a giant plant.

Two years later, Herschel retreated from this unorthodox position, but Darwin created a poetic version of Herschel's ideas, imagining that Nature arises like a phoenix from the ashes of a collapsed chaotic cosmos:

Star after star from Heaven's high arch shall rush,
Suns sink on suns, and systems systems crush,
Headlong, extinct, to one dark centre fall,
And Death and Night and Chaos mingle all!
—Till o'er the wreck, emerging from the storm,

Immortal NATURE lifts her changeful form,
Mounts from her funeral pyre on wings of flame,
And soars and shines, another and the same.[100]

During the nineteenth century, Herschel was commemorated as a great British astronomer, while Darwin was posthumously consigned to ignominious oblivion.

For many years, Darwin withheld *Zoonomia* from publication, but he eventually decided to go ahead, telling his son Robert that 'I am now too old and hardened to fear a little abuse'—every horse, he continued, must expect to be bitten by a fly.[101] Perhaps he was heartened by the government's recent decision not to prosecute William Godwin on the grounds that his controversial book was too expensive for working people to afford.[102] Stevens, the acerbic headmaster, suspected that Darwin was actually looking forward to the 'hornet-host of Assailants' he affected to fear.[103]

Progress pervades Erasmus Darwin's evolutionary vision. 'Perhaps all the productions of nature are in their progress to greater perfection?' he suggested, as if imagining that Nature is constantly striving to better herself.[104] In *Zoonomia* he wrote about 'the improving excellence observable in every part of the creation'.[105] Darwin's absentee God gave the first spark of life to matter, but then withdrew, standing back as living creatures evolved according to natural laws (although he is vague about how that might happen). For his critics, this lack of divine direction was the most controversial aspect of his ideas. He tried to protect himself against accusations of atheism by stressing that his atomic doctrine 'would strengthen the demonstration of the existence of a Deity, as the first cause of all things'.[106] However, his processes of change had no final purpose, no overall plan imposed by God, and it was this point that horrified Darwin's most influential commentator, William Paley, a wealthy philosopher and clergyman.

In 1802, the year Darwin died, Paley published his most famous book, *Natural Theology*. One image has come to summarize its central message— the watchmaker analogy. In Paley's version of this old parable, he imagines himself walking across a heath. If I stumbled over a stone, says Paley, I might assume that it had lain there forever. But if I saw a watch lying on the ground, I would know that somebody had dropped it and—crucially—that somebody had made it. Similarly, Paley continued, when we look at the living world around us, we can infer that it has been made for a purpose. God is the Great Designer, and ours is a teleological universe, one with a goal that He has planned for a reason. Paley devoted several pages of his book to damning an opponent who remained unnamed, but was easily identified by reviewers as Darwin. In Paley's eyes, Darwin represented yet another threat to Christianity at a time when it was also under attack from French Revolutionaries and British philosophers such as David Hume.[107]

Surely, argued Paley, the human eye could not possibly have arisen by chance? Although Darwin had no opportunity to respond, his grandson did. Paley regarded living organisms as complex machines deliberately made by God, but Charles Darwin eliminated divine intention, replacing it with the agency of natural selection. One great advantage of his approach is that it explains why eyes are often defective and why they wear out so quickly. According to Paley's logic, if something is useful—an eye for seeing, or a spider's web to catch a fly—then some external agent must have intended it to be that way. For him, God was the only possibility.

Charles Darwin disagreed, giving example after example to demonstrate how animal breeders could mould the living world to suit their needs by picking out desirable traits. Natural selection, he argued, operates in the same way as artificial selection. To refute his opponents' insistence that God the Grand Architect must be involved, he imagined a human architect building a house from rocks found at the bottom of a cliff. He would naturally choose the shapes that best suited their position, such as squat rectangular blocks for the foundation or tall slim ones for the door frames. These stones would be the right ones for their particular purpose, even though they had not originally been designed with that house in mind.[108]

Coincidentally, when Charles Darwin was a student at Cambridge, he lived in the same rooms that Paley had once occupied. Studying *Natural Theology* was, he wrote later, 'the only part of the academical course which, as I then felt, and as I still believe, was of the least use to me in the education of my mind'.[109] He was grappling with a book written in part to countermand the ideas of his own grandfather.

Society

For modern non-believers like me, one of the hardest aspects of Christianity to grasp is how an all-merciful God can permit the existence of evil and suffering. This quandary is technically known as theodicy, although the term was introduced only in the eighteenth century, by Gottfried Leibniz, Newton's German rival. Immersed in grief after the sudden death of his favourite pupil, the first Queen of Prussia, Leibniz struggled to understand why this young woman had suddenly ceased to exist.

Previous generations had accepted suffering as the will of God, but Enlightenment philosophers were determined to find rational explanations for everything. Leibniz's ambition was to deduce the metaphysical basis of the cosmos from a few first principles that are self-evidently true. He called one of them the Principle of Sufficient Reason: nothing—a stone, a person, an earthquake—can be the way it is without a reason, even if that reason remains unfathomable. Every single tiny event, he argued, must have an explanation. If you say that a billiard ball moves because another one hit it, then you still have to give a reason for the first one moving. Going back and back in the chain of causes, there must be one initial cause that lies outside the entire system—and that must be God. Having established that to his satisfaction, Leibniz took another step. God is perfect, and God is unique, which means that He can have no rivals. Therefore, he continued, this world must lie slightly below perfection. On the other hand, it must be as good as it possibly can be, because otherwise God would have improved it. When a beautiful young queen is suddenly struck down, you just have to accept her premature demise as a regrettable example of the inevitable defects in God's creation.

Not everybody was convinced by Leibniz's argument, and to refute it Voltaire produced what is now his best-known book: *Candide*. 'All is for the best in the best of all possible worlds,' insists Leibniz's unrealistically optimistic spokesman as disaster follows disaster. Darwin shared Voltaire's scepticism about God's intentions, struggling to reconcile his faith in progress with the undoubted misery he saw about him. He watched his first wife slowly fade away in agony, while his second son suffered from bouts of depression and committed suicide. Professionally as a doctor, he was forced time and again to admit that he could do little more than help his patients die less uncomfortably, and he came to regard the struggle for existence as a perpetual clash between good and evil, the theme of his fourth canto. After witnessing the pain of nine local children afflicted with measles, Darwin wrote to Watt that 'there is a perpetual war carried on between the Devil and all holy men'.[110]

In *The Temple of Nature*, Darwin paints dismal scenes of rapacity, misery, and death resembling the heartless 'Nature, red in tooth and claw' that has come to symbolize Charles Darwin's competition for survival (even though it originated in a poem by Alfred Lord Tennyson published years before *On the Origin of Species*). In couplet after violent couplet, wolves tear apart innocent lambs, eagles swoop on helpless doves, and sharks gobble up shoals of fish (or scaly broods, as Darwin prefers to call them—an over-poetic device known as periphrasis). For me, the most striking aspect of Darwin's grim vision is not the cruelty of these well-known predators, but the universality of natural ruthlessness. Dragonflies crunch through hordes of insects with metallic jaws, larvae explode their hosts from the inside, bees slaughter enemies by the thousand, and even plants are engaged in warfare as ivy strangles elm trees and blight decimates crops. Humanity is not exempt. When he heard that Watt had lost his only surviving daughter, Darwin responded gloomily: 'I am surprized [*sic*] that we live, rather than that our friends die.'[111] Some evils are self-inflicted—wars, slavery, drunkenness, ambition, envy—while others are due to unpredictable and unpreventable disasters such as volcanoes and famines.

Feeling worn down by all this horror, I skipped forward a few pages, and was relieved to find the poem veering towards a celebration of fecundity. Soon, Darwin once again inundates his reader with a surfeit of examples—pregnant poppies, prolific aphids, sex-crazed snails, tadpoles, herring spawn.... This teeming over-abundance builds up to a rhetorical justification of human suffering. Death, Darwin argues, is essential for preventing a population explosion that would outrun the world's resources:

> So human progenies, if unrestrain'd,
> By climate friended, and by food sustain'd,

O'er seas and soils, prolific hordes! would spread
Erelong, and deluge their terraqueous bed;
But war, and pestilence, disease, and dearth,
Sweep the superfluous myriads from the earth...[112]

Although Knight had expressed similar views, he had restricted his struggle to the animal kingdom, explaining how each species is kept in check
by the predators that devour it. Darwin's scheme was more ambitious,
because he extended the fight for existence to the human race. Significantly, he brackets together natural forces such as drought and disease
with the social evil of war, which he presents as an inevitable consequence
of civilization.[113]

Such ideas were new to Darwin and his contemporaries. In 1798—the
year of 'The Loves of the Triangles', when Darwin was working on *The
Temple of Nature*—the political economist and Anglican clergyman Thomas Malthus had published *An Essay on the Principle of Population*.
Malthus has been blamed for the rise of eugenics, the use of contracep-
tives and Sinophobic panics about 'The Yellow Peril', but these accusations stem from later editions, which were substantially revised. In this
first version, Malthus railed against the utopian visions of radicals like
Godwin, who dreamed of creating a happy, peaceful society unafflicted
by crime, disease or war. In a Malthusian world, existence without misery is impossible.[114]

Malthus starts with a quantitative argument: populations grow more
quickly than the rate at which food supplies increase. Whether it be people, plants or animals, survival is only possible if population growth is
restricted in some way. Viewed from that perspective, wars, disease and
famine provide inbuilt safety valves to cope with the world's limited capacity. Dressed up in heroic couplets, that was also Darwin's thesis. There are
no convenient letters from Darwin reporting 'Today I read Malthus's
Essay...', although it seems likely that he did; in any case, he would surely
have come across some of the many reviews that appeared. This was a
book that provoked instant controversy, and it influenced some of the
nineteenth century's most famous writers—not only Charles Darwin but
also Karl Marx and John Stuart Mill.

In addition to agreeing about the self-correcting value of wars and food shortages, Malthus and Darwin developed similar moral codes, but with different justifications. According to Malthus, God created evil to spur people into greater activity, an argument from design that strongly influenced Paley. On this line of thought, people should not let themselves be submerged in despair by wrongdoing, but instead should constantly act to overcome it and improve the life of those around them. Darwin removed the emphasis on God. Referring back to Pythagoras, he outlined a rational argument for altruism, explaining that when a plant or animal dies, its atoms will eventually be recycled. Tiny particles of matter are constantly circulating through the cosmos, so that today's lovers may contain remnants of yesterday's enemies. This physical model led Darwin to an ethical conclusion endorsing cooperation:

> While Nature sinks in Time's destructive storms,
> The wrecks of Death are but a change of forms;
> Emerging matter from the grave returns,
> Feels new desires, with new sensations burns...
> Whence drew the enlighten'd Sage the moral plan,
> That man should ever be the friend of man...[115]

Whatever uncertainty there may be about Erasmus Darwin's familiarity with Malthus, Charles Darwin was enormously affected by his ideas. Some four decades earlier, in lines Charles marked out in his personal copy of *The Temple of Nature*, his grandfather had already painted nature as a battlefield:

> —Air, earth, and ocean, to astonish'd day
> One scene of blood, one mighty tomb display!
> From Hunger's arm the shafts of Death are hurl'd,
> And one great Slaughter-House the warring world![116]

A few months after returning from his voyage on the *Beagle*, Charles Darwin spent a week at his parents' house. There he started his Notebook B, which contains his first tentative sketch of branching evolution—and on the very first page he wrote 'Zoonomia', underlining his title as if to continue his grandfather's initiative. A couple of months later, he read Malthus's *Essay*, and according to his own account, it was this book that prompted him to go still further in considering the struggle for survival in the face of competition for scarce resources. Whereas his grandfather highlighted warfare between different animals, Charles Darwin introduced the significance of fighting within the same species.

Can it be coincidence that a few days after reading Malthus, Charles Darwin yet again turned to *Zoonomia*? For Erasmus Darwin, *Zoonomia* was the prose origin of his *Temple of Nature*, but in his *Autobiography* Charles Darwin remarked that he was disappointed to find 'the proportion of speculation being so large to the facts given'.[117] The extent of *Zoonomia*'s influence on the *Origin of Species* must remain speculative, but it cannot be dismissed. Whether deliberately or unconsciously, authors often reinterpret their memories, and Charles Darwin may have leant on his grandfather's book more heavily than he was willing to acknowledge. In any case, *Zoonomia* did provide a vital medical source of information for a later controversial book by Charles Darwin—*The Expression of the Emotions in Man and Animals* (1872).[118]

CONCLUSION: REPUTATIONS AND REFLECTIONS

The trumpet of Liberty sounds through the world,
    And the universe starts at the sound ...
Shall Britons the chorus of liberty hear
    With a cold and insensible mind?
No,—the triumphs of freedom each Briton shall share,
    And contend for the rights of mankind.

John Taylor, 'The Trumpet of Liberty',
Norfolk Chronicle (16 July 1791)

I first encountered Erasmus Darwin long ago, when I picked up Mary Shelley's *Frankenstein* (1818) and read the opening sentence of her husband's preface: 'The event on which this fiction is founded has been supposed, by Dr Darwin, and some of the physiological writers of Germany, as not of impossible occurrence.'[1] Darwin's reputation was then, it seems, still sufficiently high to warrant citing him as an authority figure.

Frankenstein was written by a teenager, but the original novel is far more intriguing than the Boris Karloff movie version. Years later, when she was in her mid-thirties, Shelley thought back to the genesis of her story, and recalled rumours that Darwin had brought a piece of vermicelli to life. Presumably she was referring not to a type of pasta, but to the word's literal meaning—'small worms' (that might sound an irrelevant detail, but she's often mocked for getting it wrong). What Darwin actually reported was that a vorticella, a minute organism found in stagnant water, can be soaked back into activity after several months of dryness.[2] Appropriate, perhaps, that in 1878, Darwin's coffin burst open to reveal him dressed in his purple velvet dressing gown, perfectly preserved, his features unaltered.[3]

Finished works conceal the processes involved in creating them. When Mary Shelley started *Frankenstein*, the first words she set down on paper were 'It was a dark and stormy night ...'; in the course of drafting and redrafting, and with many interventions by Percy Shelley, this opening moved first to chapter seven and eventually ended up in chapter four. On the other hand, the reference to Dr Darwin that appeared at the beginning was added only at the last minute, when the manuscript was finally ready for the publishers.

Shelley was right to start in the middle (at least, that's how I work). For one thing, she avoided the blocked writer's panic of confronting a blank page, when the first sentence acquires such significance that it never reaches the paper. Far better to set down something that is later abandoned than to remain trapped in indecision, straining after unattainable perfection. It is only through writing that you know what you want to say and, still more fundamentally, what questions you want to ask. Explaining makes you think, which is why you need to start getting those words down, long before you feel you have something definitive to report. Editing comes later, and the final product will be far removed from those preliminary attempts.

Academic conventions require authors to give the impression of having progressed steadily in a straight line from a well-formulated question to an incontrovertible answer. In reality, attempts to follow such an ideal path are doomed to failure. Setbacks, false trails, and jettisoned chapters are not only inevitable but also essential to discovery. Like anyone drawing towards the end of a project, I am now a different person. If I were to begin again, my meandering route towards Erasmus Darwin would start from a different place. A few years ago, when I knew even less than I do now, I included these two paragraphs in my original proposal for this book:

> *Airbrushed out of history during the nineteenth century, Darwin started receiving accolades from scientists in the last third of the following one. In particular, the geomagnetist Desmond King-Hele waged what was initially*

a one-man campaign to arouse interest in Charles Darwin's neglected grand-father. It is thanks to him that so many biographical details are now readily available, and that Darwin's house in Lichfield has been restored. King-Hele resurrected Darwin as an unjustly forgotten genius who not only solved major evolutionary puzzles but also scored high marks for anticipating later discoveries—the steering mechanism used in modern cars, the gas laws of clouds, internal combustion engines fuelled by hydrogen, alarm clocks, sewing machines and more. As if that were not enough, King-Hele awarded Darwin points for his strong influence on Romanticism, although regrettably, English literature specialists still exclude him from their anthologies.

Other versions of Darwin have also appeared, tailored to suit the ideological agendas of different biographers. Initially condemned by obituarists as a pedantic poet who leant towards Venus rather than God, by 1990 Darwin had been denounced as an oppressive male chauvinist and an exploitative capitalist. Whether intentionally or not—scholars argued—Darwin presented in his poetry the privileged perspective of a well-off, well-educated patriarch. Although ostensibly writing to enlighten women, Darwin was establishing a masculine botany designed to titillate gentlemen and reinforce their superior status by converting female physiques and psyches into stereotypes. As the nation's poetic champion of industrial progress, Darwin celebrated the technological advances that boosted production rates and profits, but—like artists of the time—he portrayed utopian landscapes empty of labourers. When appraised retrospectively, with knowledge of the Victorian slums and child labour that followed, Darwin appears to have been dazzled by the benefits of technical innovation but blind to its destructive potential.

There is no 'true' version of Darwin or, indeed, of any other character from the past, but I have aimed to complement rather than to supplant the previous interpretations on which I based my proposal.

Like all historians, I too have a vantage point. In writing this book, I have been affected not only by current academic interest in slavery, ethnicity, and the Darwin family, but also by my own life. As the daughter of a Jewish German immigrant, and the mother-in-law of a black American, I feel involved in current debates about apologizing for the past. Personal experiences shape your academic research—and the influence also works in the opposite direction. Thinking about history alters how you behave and what your hopes are for the future.

As I wrote and learnt more, my plans altered. When I began, I was intrigued by a 'what if?...' type of question. If Erasmus Darwin had not had a famous grandson, I wondered, would anyone nowadays show any interest in him? Of course, his position as a doctor publishing best-selling poems made him unique, but from his own perspective, he was—like everybody else—struggling through life and trying to make the best that he could out of it. As was the case for everyone, what he did and thought was affected by personal events—deaths, romances, children—as well as by public ones, such as the revolutions in Paris and Haiti, or the ruthless slaughter of slaves thrown off the *Zong*.

After several years of immersion in Darwin's writing, I still have a low opinion of his poetic skills. On the other hand, I have come to admire his passionate commitment to making the world a better place. He must have been a marvellously invigorating yet infuriating man to know. Through his medical treatments, mechanical inventions, and enthusiastic support of the Lunar Society, Darwin took practical steps to help others and improve Britain. In turning to poetry, he did not simply step aside from his previous path through life, but began trying to achieve his ideals in a different way. As someone who abhorred suffering, whether physical or mental, Darwin wanted to change how society operates. The *Anti-Jacobin* satirists, on the other hand, were wedded through self-interest to stability. They set about undermining his political mission by attacking him at his weakest point—his heroic couplets—and so encouraged critics to ignore the substance of his arguments.

In his portraits, Darwin resembles many other overweight eighteenth-century gentlemen in a wig. He looks pompous and strait-laced, very different from the exuberant, sensitive, and self-doubting physician I had come to appreciate from his correspondence and his poetry. Unlike the Lunar Society industrialists who stressed the value of automation, Darwin was fascinated by life, by the inner drives and energies that enable animals to experience pleasure, to adapt their environments, to alter their very nature. For him, sexual reproduction offered not only great joy to living creatures (including himself), but also the opportunity for species to improve. Benefiting from the prehensile hands and inventive minds inherited from their animal predecessors, his human beings can create powerful machinery. They cannot, however, be reduced to machine-like organisms.[4]

I tried to escape from the confines of my own discipline, the history of science, by setting Darwin within a non-scientific context. Instead of focusing exclusively on his industrial innovations, medical treatments, and botanic translations, I read his love letters and compared him with all those other would-be iconic poets who never acquired the stature of Pope or Dryden, of Wordsworth or Coleridge. Darwin was far from alone in writing didactic, footnoted poems, which—however tedious they might seem to me—enjoyed great popularity in the eighteenth century. In particular, thanks to the inspiring work of Martin Priestman, a professor of English literature, I learnt to appreciate Darwin as a Lucretian poet. Instead of grouping him with Linnaeus, Lamarck, and the Lunar men, I bracketed him with Richard Payne Knight and William Jones. All three combined botany, mythology, and social commentary in their poetry, and all three favoured developmental, evolutionary models of nature, language, and society.

Alongside these unanticipated insights came disappointments. When I started out, I hoped to find evidence that Darwin's opposition to slavery powerfully influenced his ideas about evolution, but I was forced to abandon such an optimistically neat correlation. Abolition and evolution were certainly consistent with each other, but that is not the same as proof of a causal relationship. I found little evidence that abolitionists relied on scientific arguments rather than religious, moral, or economic ones.

On the other hand, I did conclude that there was a less direct relationship between Darwin, abolition, and evolution. Through his poetry, Darwin publicized the anti-slavery logo devised by his close friend Wedgwood. The grandson these two men held in common—Charles Darwin—was brought up in a family that inherited and passed down this antipathy. By the time Charles Darwin was an adult, scientific theories had brought race into existence as a meaningful category. Like his two grandfathers, he responded angrily to slavery, and this egalitarian outlook affected his evolutionary hypotheses. Appalled at seeing African captives in Brazil, he fumed that black slaves 'are ranked by the polished savages in England as hardly their brethren, even in Gods eyes'. His choice of words recalls his grandparents' slogan: 'Am I not a man and a brother?'[5]

Gradually I realized that what I really wanted to do was make sense of 'The Loves of the Triangles'. To be honest, the poem first appealed to me as a gap in the scholarly market, a satire that was often referred to but only scantily discussed—ideal for a research topic. But as so often happens, the more I studied it, the more fascinating it seemed, and the deeper I became involved in deciphering its complex puns and references. I am resigned to accepting that many of its subtleties will forever pass me by, but the parts I did manage to understand revealed political aspects of Darwin that had not been examined before.

Politicizing Darwin by no means negates his scientific significance, but does entail thinking about science and technology from fresh angles. Consider the Lunar Society, now regarded as crucial for the development of British manufacturing. For historians living after the era of Victorian industrialization, it makes sense to characterize the Lunar network in this way. In contrast, the participants themselves attached so much significance to their political allegiances that the Society fell apart after Priestley's house was destroyed by rioters and he was branded a Jacobin supporter. I gave a different spin to the Lunar Society by focusing on those members who were active social reformers. In this revisionary version—which is no more the 'right' one than the conventional scientific description—Thomas Day and Josiah Wedgwood feature as prominent abolitionists, Richard Edgeworth and Erasmus Darwin as campaigners for female education, and Joseph Priestley as an outspoken political radical. Under this interpretation, two members who were valued at the time but are rarely mentioned by historians—William Small and John Whitehurst—gain new significance as open supporters of the American Revolution.

Since Darwin's contemporaries paid him so much serious attention, so too should historians. The 'Loves of the Triangles' was no mere jibe at an overblown semi-erotic poem. It was a meticulously crafted piece of political propaganda based on a close study of Darwin's work—and the parodists assumed that readers would pick up on all those references that now seem so arcane. Darwin's poetry was deeply political—in an important sense, even more so than the *Anti-Jacobin*'s. His parodists focused on immediate and local concerns of revolution and republicanism: although their satire includes a stunningly wide range of references, it is preoccupied with specific events of the immediate past and directed against contemporary figures. Darwin's scope is broader. Especially in his numerous references to his Lunar colleagues, he does raise issues of his age, but he places them within a general thesis of the human condition—and he sets that within a historical sketch of the universe.[6]

The poet mocked for being shallow was more profound than his *Anti-Jacobin* critics. 'The Loves of the Triangles' illustrates how easy it was and is to mock Erasmus Darwin. He is vulnerable to satire on several fronts— his ponderous heroic couplets, his anthropomorphic portrayals of plants, his vague references to filaments and heat, his enthusiastic descriptions of subterranean gnomes and long-distance balloon flights. Delightful to parody because he wrote so ineptly, Darwin undermined his own profundity and novelty by relying on an inappropriate old-fashioned poetic convention to express the ideas of Romanticism.

By pursuing Darwin through the verses of the *Anti-Jacobin*, I have tried to appraise him and his poetry with the eyes of his contemporaries. One consequence of being conducted by serendipity has been to trace an unfamiliar and erratic path through the terrain of the eighteenth century, one that meanders towards an evolutionary future but originates in classical culture. I hope that this eclectic route has made me an entertaining as well as an informative travel guide—and that, like me, you have come to appreciate the thoughtful, energetic, and witty man who lies concealed behind his poetry.

APPENDIX
'The Loves of the Triangles'

This version is transcribed from *Poetry of the Anti-Jacobin. The fifth edition* (London, 1807). The superscripts in Arabic numerals correspond to the authors' original footnotes; the superscripts in lower-case Roman letters refer to my own editorial comments, which are at the end of the poem as endnotes.

In order to maintain the appearance and reading experience of contemporary readers, I have incorporated the satirists' footnotes in a similar but computerized style to their own. Although in the original version there were no superscripts in the text, the footnotes did appear at the foot of each page, so that they could be read in conjunction with the verse. This was standard eighteenth-century practice, and imitates the style of Erasmus Darwin himself.

The authors' footnoting style changed a third of the way through. In the first two instalments, they made no mark in the verse, but indicated the line number in the note; for those, I have put the note number at the end of the relevant line. In the last instalment, they changed to using asterisks and other printers' marks to indicate a note; I have replaced those symbols by footnote numbers.

The first instalment was by Canning and Frere; the second by Canning, Gifford and Frere; and the third by Canning.

My editorial end-notes offer brief guidance but are not intended to be comprehensive. This is partly through ignorance, but I have also intentionally not duplicated points I have elaborated more fully in the main text of the book.

No. XXIII.

April 16.

We cannot better explain to our Readers, the design of the Poem from which the following Extracts are taken, than by borrowing the expressions of the Author, Mr. Higgins, of St. Mary Axe, in the letter which accompanied the manuscript.[i]

We must premise, that we had found ourselves called upon to remonstrate with Mr. H. on the freedom of some of the positions laid down in his other Didactic Poem, the Progress of Man;[ii] and had in the course of our remonstrance, hinted something to the disadvantage of the *new principles* which are now afloat in the world; and which are, in our opinion, working to much prejudice to the happiness of mankind. To this Mr. H. takes occasion to reply—

"What you call the *new principles* are, in fact, nothing less than *new.* They are the principles of primeval nature, the system of original and unadulterated man.

"If you mean by my addiction to *new principles,* that the object which I have in view in my larger Work (meaning the Progress of Man) and in the several other *concomitant* and *subsidiary* Didactic Poems which are necessary to complete my plan, is to restore this first and pure simplicity; to rescue and recover the interesting nakedness of human nature, by ridding her of the cumbrous establishments which the folly, and pride, and self-interest of the worst part of our species have heaped upon her;—you are right.—Such is my object. I do not disavow it. Nor is it mine alone. There are abundance of abler hands at work upon it. *Encyclopedias, Treatises, Novels, Magazines, Reviews,* and *New Annual Registers,* have, as you are well aware, done their part with activity and with effect. It remained to bring the *heavy* artillery of a Didactic Poem, to bear upon the same object.

"If I have selected your Paper as the channel for conveying my labours to the Public, it was not because I was unaware of the hostility of your principles to mine, of the bigotry of your attachment to 'things as they are:'—but because, I will fairly own, I found some sort of cover and disguise necessary for securing the favourable reception of my sentiments; the usual pretexts of humanity and philanthropy, and fine feeling, by which we have for some time obtained a passport to the hearts and under-standings of men, being now worn out, or exploded. I could not choose but smile at my success in the first instance in inducing *you* to adopt my Poem as your own.

"But you have called for an explanation of these principles of ours, and you have a right to obtain it. Our first principle is, then—the reverse of the trite and dull maxim of Pope—'*Whatever is, is right.*'iii We contend, that '*Whatever is, is wrong:*'—that Institutions civil and religious, that Social Order (as it is called in *your* cant) and regular Government, and Law, and I know not what other fantastic inventions, are but so many cramps and fetters on the free agency of man's *natural intellect* and *moral sensibility*; so many badges of his degradation from the primal purity and excellence of his nature.

"Our second principle is the '*eternal and absolute Perfectibility of Man.*'iv We contend, that if, as is demonstrable, we have risen from a level with the *cabbages of the field* to our present comparatively intelligent and dignified state of existence, by the mere exertion of our own *energies;* we should, if these *energies* were not repressed and subdued by the operation of preju-dice, and folly, by KING-CRAFT and PRIEST-CRAFT, and the other evils incident to what is called Civilized Society, continue to exert and expand ourselves in a proportion infinitely greater than any thing of which we yet have any notion: in a *ratio* hardly capable of being calculated by any sci-ence of which we are now masters; but which would in time raise Man from his present biped state to a rank more worthy of his endowments and aspirations; to a rank in which he would be, as it were, *all* MIND; would enjoy unclouded perspicacity and perpetual vitality; feed on *Oxygene,* and never die, but *by his own consent.*v

"But though the Poem of the PROGRESS OF MAN alone would be suf-ficient to teach this system, and enforce these doctrines; the whole practi-cal effect of them cannot be expected to be produced but by the gradual perfecting of each of the sublimer sciences;—at the husk and shell of which we are now nibbling, and at the kernel whereof, in our present state, we cannot hope to arrive. These several Sciences will be the subjects of the several *auxiliary* DIDACTIC POEMS which I have now in hand (one of which, entitled THE LOVES OF THE TRIANGLES, I herewith transmit to you) and for the better arrangement and execution of which, I beseech you to direct your Bookseller to furnish me with a handsome Chambers's Dictionary; in order that I may be enabled to go through, the several articles alphabetically, beginning with *Abracadabra,* under the first letter, and going down to *Zodiac,* which is to be found under the last.

"I am persuaded that there is no Science, however abstruse, nay, no Trade or Manufacture, which may not be taught by a Didactic Poem. In that before you, an attempt is made (not unsuccessfully I hope) to *enlist the Imagination under the banners of Geometry.*[vi] *Botany* I found done to my hands. Although the more rigid and unbending stiffness of a mathematical subject does not admit of the same appeals to the warmer passions, which naturally arise out of the *sexual* (or, as I have heard several worthy Gentlewomen of my acquaintance, who delight much in the Poem to which I allude, term it, by a slight misnomer no way difficult to be accounted for— the *sensual*) system of Linnaeus;[vii]—yet I trust that the range and variety of illustration with which I have endeavoured to ornament and enlighten the arid truths of Euclid and Algebra, will be found to have smoothed the road of Demonstration, to have softened the rugged features of Elementary Propositions, and, as it were, to have strewed the *Asses' Bridge* with flowers."[viii]

Such is the account which Mr. HIGGINS gives of his own undertaking, and of the motives which have led him to it. For our parts, though we have not the same sanguine persuasion of the *absolute perfectibility* of our species, and are in truth liable to the imputation of being more satisfied with *things as they are,* than Mr. HIGGINS and his Associates;—yet, as we are in at least the same proportion, less convinced of the practical influence of Didactic Poems, we apprehend little danger to our Readers' morals from laying before them Mr. HIGGINS's doctrine in its most fascinating shape. The Poem abounds, indeed, with beauties of the most striking kind— various and vivid imagery, bold and unsparing impersonifications; and similitudes and illustrations brought from the most ordinary and the most extraordinary occurrences of nature,—from history and fable,—appealing equally to the heart and to the understanding, and calculated to make the subject of which the Poem professes to treat, rather amusing than intelligible. We shall be agreeably surprised to hear that it has assisted any young Student, at either University, in his Mathematical Studies.

We need hardly add, that the Plates illustrative of this poem (the engravings of which would have been too expensive for our publication) are to be found in Euclid's Elements, and other books of similar tendency.

LOVES OF THE TRIANGLES.
ARGUMENT OF THE FIRST CANTO.

Warning to the Profane not to approach—Nymphs and Deities of Mathematical Mythology—Cyclois of a pensive turn—Pendulums, on the contrary, playful—and why?—Sentimental union of the Naiads and Hydrostatics—Marriage of Euclid and Algebra.—Pulley the emblem of Mechanics—Optics of a licentious Disposition—distinguished by her Telescope and Green Spectacles.—Hyde Park Gate on a Sunday Morning—Cockneys—Coaches.—Didactic Poetry—Nonsensia—Love delights in Angles or Corners—Theory of Fluxions explained[ix]*—Trochais,*[x] *the Nymph of the Wheel—Smoke-Jack described—Personification of elementary or culinary Fire. —Little Jack Horner—Story of Cinderella—Rectangle, a Magician, educated by Plato and Menecmus—in love with Three Curves, at the same time—served by Gins, or Genii—transforms himself into a Cone—The Three Curves requite his Passion—description of them—Parabola, Hyperbola, and Ellipsis—Asymptotes—Conjugated Axes. —Illustrations—Rewbell, Barras, and Lepaux,*[xi] *the Three virtuous Directors—Macbeth and the Three Witches—The Three Fates—The Three Graces—King Lear and his Three Daughters—Derby Diligence*[xii]*—Catherine Wheel. —Catastrophe of Mr. Gingham,*[xiii] *with his Wife and Three Daughters overturned in One-Horse Chaise—Dislocation and Confusion, two kindred Fiends—Mail Coaches—Exhortation to Drivers to be careful—Genius of the Post Office—Invention of Letters—Digamma*[xiv]*—Double Letters-remarkable Direction of one—Hippona*[xv] *the Goddess of Hack-horses—Parameter and Abscissa unite to overpower the Ordinate, who retreats down the Axis Major, and forms himself in a Square—Isosceles, a Giant—Dr. Rhomboides—Fifth Proposition, or Asses' Bridge*[xvi]*—Bridge of Lodi*[xvii]*—Buonaparte—Raft and Windmills—Exhortation to the recovery of our Freedom—Conclusion.*

THE LOVES OF THE TRIANGLES
A MATHEMATICAL AND PHILOSOPHICAL POEM.

INSCRIBED TO DR. DARWIN.

CANTO I.

STAY your rude steps, or e'er your feet invade
The Muses' haunts, ye Sons of War and Trade!
Nor you, ye Legion Fiends of Church and Law,
Pollute these pages with unhallow'd paw![1]
Debased, corrupted, groveling, and confined,
No DEFINITIONS touch *your* senseless mind;[2]
To *you* no POSTULATES prefer their claim,[3]
No ardent AXIOMS *your* dull souls inflame;[4]
For *you* no TANGENTS touch, no ANGLES meet,[5]
No CIRCLES join in osculation[xviii] sweet![6]   10

For *me*, ye CISSOIDS, round my temples bend[7]
Your wandering Curves; ye CONCHOIDS extend;[8]

---

[1] *Ver.* 1-4. Imitated from the introductory couplet to the ECONOMY OF VEGETATION.

"Stay your rude steps, whose throbbing breasts infold
"The Legion Fiends of Glory and of Gold."
This sentiment is here expanded into four lines.

[2] Ver. 6. *Definition*—A distinct notion explaining the Genesis of a thing—*Wolfius.*

[3] Ver. 7. *Postulate*—A self-evident proposition.

[4] Ver. 8. *Axiom*—An indemonstrable truth.

[5] Ver. 9. *Tangents*—*So* called from touching, because they touch Circles, and never cut them.

[6] Ver. 10. *Circles*—*See* Chambers's Dictionary—Article Circle. Ditto. *Osculation*—For the *Osculation*, or kissing of Circles and other Curves, see Huygens, who has veiled this delicate and inflammatory subject in the decent obscurity of a learned language.

[7] Ver. 11. *Cissois*—A Curve supposed to resemble the sprig of ivy from which it has its name, and therefore peculiarly adapted to poetry.

[8] Ver. 12. *Conchois,* or *Conchylis*—A most beautiful and picturesque Curve; it bears a fanciful resemblance to a *Conch* shell. The Conchois is capable of infinite extension, and presents a striking analogy between the Animal and Mathematical Creation. Every individual of this species, containing within itself a series of *young* Conchoids for several generations, in the same manner as the Aphides, and other insect tribes, are observed to do.

Let playful PENDULES quick vibration feel,
While silent CYCLOIS rests upon her wheel;
Let HYDROSTATICS, simpering as they go,[9]   15
Lead the light Naiads on fantastic toe;
Let shrill ACOUSTICS tune the tiny lyre;[10]
With EUCLID sage fair ALGEBRA conspire;[11]
The obedient pulley strong MECHANICS ply,[12]
And wanton OPTICS roll the melting eye!   20

I see the fair fantastic forms appear,
The flaunting drapery and the languid leer;
Fair Sylphish forms—who, tall, erect, and slim,[13]
Dart the keen glance, and stretch the length of limb;
To viewless harpings weave the meanless dance,   25
Wave the gay wreath, and titter as they prance.

Such rich confusion charms the ravish'd sight,[14]
When vernal Sabbaths to the Park[xix] invite;
Mounts the thick dust, the coaches crowd along,
Presses round Grosvenor Gate the impatient throng;   30
White-muslin'd misses and mammas are seen,
Link'd with gay Cockneys glittering o'er the green:
The rising breeze unnumber'd charms displays,
And the tight ankle strikes the astonish'd gaze.

But chief, thou Nurse of the Didactic Muse,   35
Divine NONSENSIA, all thy sense infuse;

---

[9] Ver. 15. *Hydrostatics*—Water has been supposed, by several of our philosophers, to be capable of the passion of Love.—Some later experiments appear to favour this idea—Water, when pressed by a moderate degree of heat, has been observed to *simper,* or *simmer* (as it is more usually called.) —The same does not hold true of any other element.

[10] Ver. 17. —*Acoustics*—The doctrine or theory of sound.

[11] Ver. 18. —*Euclid and Algebra*—The loves and nuptials of these two interesting personages, forming a considerable Episode in the Third Canto, are purposely omitted here.

[12] Ver. 19. *Pulley*—So called from our Saxon word, to PULL, signifying to pull or draw.

[13] Ver. 23. *Fair Sylphish Forms*—Vide modern prints of nymphs and shepherds dancing to nothing at all.

[14] Ver. 27. *Such rich confusion*—Imitated from the following genteel and sprightly lines in the First Canto of the LOVES OF THE PLANTS :-

> So bright its folding canopy withdrawn,
> Glides the gilt landau o'er the velvet lawn,
> Of beaux and belles displays the glittering throng,
> And soft airs fan them as they glide along.

The charms of *Secants* and of *Tangents* tell,
How Loves and Graces in an *Angle*ˣˣ dwell;[15]
How slow progressive *Points* protract the *Line*,[16]
As pendant spiders spin the filmy twine;   40

   How lengthen'd *Lines,* impetuous sweeping round,
Spread the wide *Plane,* and mark its circling bound;
How *Planes,* their substance with their motion grown,
Form the huge *Cube,* the *Cylinder,* the *Cone.*

---

[15] Ver. 38. *Angle*—Gratus puellæ risus ab Angulo. —*Hor.*
[16] Ver. 39. *How slow progressive Points*—The Author has reserved the picturesque imagery which the *Theory of Fluxions* naturally suggested for his ALGEBRAIC GARDEN; where the *Fluents* are described as rolling with an even current between a margin of *Curves* of the higher order, over a pebbly channel, inlaid with *Differential Calculi.*

In the following six lines he has confined himself to a strict explanation of the Theory, according to which Lines are supposed to be generated by the motion of Points; —Planes by the lateral motion of Lines; —and Solids from Planes, by a similar process.

Quaere—Whether a practical application of this Theory would not enable us to account for the Genesis, or original formation of Space itself, in the same manner in which Dr. Darwin has traced the whole of the organized creation to his Six Filaments—Vide ZOONOMIA. We may conceive the whole of our present Universe to have been originally concentered in a single Point. —We may conceive this Primeval Point, or *Punctum Saliens* of the Universe, evolving itself by its own energies, to have moved forward in a right Line, *ad infinitum,* till it grew tired—After which, the right Line, which it had generated, would begin to put itself in motion in a lateral direction, describing an Area of infinite extent. This Area, as soon as it became conscious of its own existence, would begin to ascend or descend, according as its specific gravity might determine it, forming an immense solid space filled with Vacuum, and capable of containing the present existing Universe.

Space being thus obtained, and presenting a suitable Nidus, or receptacle, for the generation of Chaotic Matter, an immense deposit of it would gradually be accumulated: —After which, the Filament of *Fire* being produced in the Chaotic Mass, by an *Idiosyncracy,* or self-formed habit, analogous to fermentation, *Explosion* would take place; *Suns* would be shot from the Central Chaos; —*Planets* from *Suns;* and *Satellites* from *Planets.* In this state of things, the Filament of *Organization* would begin to exert itself in those independent masses which, in proportion to their bulk, exposed the greatest surface to the action of *Light* and *Heat.* This Filament, after an infinite series of ages, would begin to *ramify,* and its viviparous offspring would diversify their forms and habits, so as to accommodate themselves to the various *incunabula* which Nature had prepared for them. —Upon this view of things, it seems highly probable that the first effort of Nature terminated in the production of Vegetables, and that these being abandoned to their own *energies,* by degrees detached themselves from the surface of the earth, and supplied themselves with wings or feet, according as their different propensities determined them in favour of aerial and terrestrial existence. Others by an inherent disposition to society and civilization, and by a stronger effort of *volition,* would become Men. These, in time, would restrict themselves to the use of their *hind feet:* their *tails* would gradually rub off by sitting in their caves or huts, as soon as they arrived at a domesticated state: they would invent *language,* and the use of *fire,* with our present and hitherto imperfect system of *Society.* In the mean while, the *Fuci* and *Algæ,* with the *Corallines* and *Madrepores,* would transform themselves into *fish,* and would gradually populate all the sub-marine portion of the globe.

Lo! where the chimney's sooty tube ascends,   45
The fair TROCHAIS[xxi] from the corner bends![17]
Her coal-black eyes upturn'd, incessant mark
The eddying smoke, quick flame, and volant spark;
Mark with quick ken, where flashing in between,
Her much loved *Smoke-Jack* glimmers thro' the scene;   50
Mark how his various parts together tend
Point to one purpose, —in one object end:
The spiral *grooves* in smooth meanders flow,
Drags the long *chain,* the polish'd axles glow,
While slowly circumvolves the piece of beef below: 55
The conscious fire with bickering radiance burns,[18]
Eyes the rich joint, and roasts it as it turns.
So youthful Horner roll'd the roguish eye,
Cull'd the dark plum from out his Christmas pye,
And cried in self-applause— "How good a boy am I !"   60

So she, sad victim of domestic spite,
Fair Cinderella, past the wintry night,
In the lone chimney's darksome nook immured,
Her form disfigured, and her charms obscured.
Sudden her God-mother appears in sight,   65
Lifts the charm'd rod, and chants the mystic rite.
The chanted rite the maid attentive hears,
And feels new ear-rings deck her listening ears;[19]
While 'midst her towering tresses, aptly set,
Shines bright with quivering glance, the smart aigrette;   70
Brocaded silks the splendid dress complete,

[17] Ver. 46. *Trochais—The* Nymph of the Wheel, supposed to be in love with Smoke-Jack.
[18] Ver. 56. *The Conscious Fire—*The Sylphs and Genii of the different elements have a variety of innocent occupations assigned them: those of fire are supposed to divert themselves with writing *Kunkel* in phosphorus. —See ECONOMY OF VEGETATION.

> "Or mark with shining letters Kunkel's name
> "In the slow *phosphor's* self-consuming flame."

[19] Ver. 68. *Listening ears—*Listening, and therefore peculiarly suited to a pair of diamond ear-rings. See the description of Nebuchadnezzar in his transformed state.

> Nor flattery's self can pierce his *pendant ears.*

In poetical diction, a person is said to '*breathe the* BLUE *air,*' and to '*drink the* HOARSE *wave*!' —not that the colour of the sky, or the noise of the water, has any reference to drinking or breathing, but because the Poet obtains the advantage of thus describing his subject under a *double relation,* in the same manner in which material objects present themselves to our different senses at the same time.

And the Glass Slipper grasps her fairy feet;
Six cock-tail'd mice[xxii] transport her to the ball,[20]
And liveried lizards wait upon her call.

Alas! that partial Science should approve   75
The sly RECTANGLE's too licentious love!
For *three* bright nymphs, &c. &c.

[*To be continued*][xxiii]

No. XXIV.

April 23.

THE LOVES OF THE TRIANGLES.

A MATHEMATICAL AND PHILOSOPHICAL POEM.

[*Continued.*]

CANTO I.

ALAS! that partial Science should approve   75
The sly RECTANGLE's too licentious love![21]

---

[20] Ver. 73. *Cock-tail'd mice*—coctilibus muris. *Ovid.* —There is reason to believe, that the *murine,* or *mouse* species, were anciently much more numerous than at the present day. It appears from the sequel of the line, that Semiramis surrounded the city of Babylon with a number of these animals.

> *Dicitur altam*
> *Coctilibus Muris cinxisse Semiramis urbem.*

It is not easy at present to form any conjecture with respect to the end, whether of ornament or defence, which they could be supposed to answer. I should be inclined to believe that in this instance the mice were dead, and that so vast a collection of them must have been furnished by way of tribute, to free the country from these destructive animals. This superabundance of the *murine* race must have been owing to their immense fecundity, and to the comparatively tardy reproduction of the *feline* species. The traces of this disproportion are to be found in the early history of every country. —The ancient laws of Wales estimate a Cat at the price of as much corn as would be sufficient to cover her, if she were suspended by the tail with her fore-feet touching the ground. —See Howel Dha. —In Germany, it is recorded that an army of rats, a larger animal of the *mus* tribe, were employed as the Ministers of Divine Vengeance against a feudal Tyrant; and the commercial legend of our own Whittington, might probably be traced to an equally authentic origin.

[21] Ver. 76. *Rectangle*—"A figure which has one Angle, or more, of ninety degrees." —Johnson's Dictionary. It here means a right-angled Triangle, which is therefore incapable of having more than one Angle of ninety degrees, but which may, according to our Author's *Prosopopœia,* be supposed to be in love with Three, or any greater number of nymphs.

For *three* bright nymphs the wily wizard burns;—
Three bright-ey'd nymphs requite his flame by turns.
Strange force of magic skill! combined of yore
With Plato's science and Menecmus'[xxiv] lore.[22]   80

   In *Afric's* schools, amid those sultry sands
High on its base where Pompey's pillar[xxv] stands,
This learnt the Seer; and learnt, alas! too well,
Each scribbled talisman, and smoky spell:
What mutter'd charms, what soul-subduing arts   85
Fell Zatanai to his sons imparts.[23]

   GINS[24]—black and huge! who in Dom-Daniel's[25] cave
Writhe your scorch'd limbs on sulphur's azure wave,[26]
Or, shivering, yell amidst eternal snows,
Where cloud-capp'd Caf protrudes his granite toes;[27]   90
(Bound by his will, *Judæa's* fabled king,[28]
Lord of *Aladdin's* Lamp and mystic Ring.)
Gins! ye remember! —for your toil convey'd
Whate'er of drugs the powerful charm could aid;

[22] Ver. 80. *Plato's and Menecmus' lore*—Proclus attributes the discovery of the Conic Sections to Plato, but obscurely. Eratosthenes seems to adjudge it to Menecmus "*Neque Menecmeos necesse erit in Cono secare tenarios.*" (Vide *Montucla*.) From Greece they were carried to Alexandria, where (according to our Author's beautiful fiction) *Rectangle* either did or might learn magic.

[23] Ver. 86. *Zatanai*—Supposed to be the same with Satan. —Vide the *New Arabian Nights,* translated by Cazotte, author of "*Le Diable amoreux*"

[24] Ver. 87. *Gins*—the Eastern name for Genii. —Vide Tales of ditto.

[25] Ver. 87. *Dom-Danie*—a sub-marine palace near Tunis, where Zatanai usually held his court. —Vide *New Arabian Nights.*

[26] Ver. 88. *Sulphur*—A substance which, when cold, reflects the yellow rays, and is therefore said to be yellow. When raised to a temperature at which it *attracts oxygene* (a process usually called *burning*), it emits a blue flame. This may be beautifully exemplified, and at a moderate expense, by igniting those *fasciculi* of brimstone *matches,* frequently sold (so frequently, indeed, as to form one of the London cries) by women of an advanced age, in this metropolis. They will be found to yield an *azure,* or blue light.

[27] Ver. 90. *Caf*—the Indian *Caucasus.* —Vide *Bailly's Lettres sur l'Atlantide,* in which he proves that this was the native country of Gog and Magog (now resident in Guildhall), as well as of the Peris, or fairies, of the Asiatic Romances.

[28] Ver. 91. *Judæa's fabled king*—Mr. HIGGINS does not mean to deny that Solomon was really king of Judæa. The epithet fabled, applies to that empire over the Genii, which the retrospective generosity of the Arabian fabulists has bestowed upon this monarch.

Air, earth, and sea ye search'd, and where below   95
Flame embryo lavas, young volcanoes glow, —[29]
Gins! ye beheld appall'd the enchanter's hand
Wave in dark air the *Hypothenusal* wand;
Saw him the mystic *Circle* trace, and wheel
With head erect, and far-extended heel;[30]   100
Saw him, with speed that mock'd the dazzled eye,
Self-whirl'd, in quick gyrations eddying fly:
Till done the potent spell—behold him grown
Fair *Venus'* emblem—the *Phœnician* CONE.[31]

Triumphs the Seer, and now secure observes   105
The kindling passions of the *rival* CURVES.

And first, the fair PARABOLA behold,[32]
Her timid arms with virgin blush unfold!
Though, on one *focus* fix'd, her eyes betray
A heart that glows with love's resistless sway,   110
Though, climbing oft, she strive with bolder grace
Round his tall neck to clasp her fond embrace,

[29] Ver. 96. *Young volcanoes*—The genesis of burning mountain was never, till lately, well explained. Those with which we are best acquainted are certainly not viviparous; it is therefore probable, that there exists in the centre of the earth a considerable reservoir of their eggs, which, during the obstetrical convulsions of general earthquakes, produce new volcanoes.

[30] Ver. 100. *Far extended heel*—The personification of *Rectangle*, besides answering a poetical purpose, was necessary to illustrate Mr. HIGGINS's philosophical opinions. The ancient mathematicians conceived that a Cone was generated by the revolution of a Triangle; but this, as our Author justly observes, would be impossible, without supposing in the Triangle that *expansive nisus*, discovered by Blumenbach, and improved by Darwin, which is peculiar to animated matter, and which alone explains the whole mystery of organization. Our enchanter sits on the ground, with his heels stretched out, his head erect, his wand (or *Hypothenuse*) resting on the extremities of his feet and the tip of his nose, (as is finely expressed in the engraving in the original work), and revolves upon his bottom with great velocity. His skin, by magical means, has acquired an indefinite power of expansion, as well as that of assimilating to itself all the *azote* of the air which he decomposes by expiration from his lungs—an immense quantity, and which in our present unimproved and uneconomical mode of breathing, is quite thrown away. By this simple process the transformation is very naturally accounted for.

[31] Ver. 104. *Phœnician Cone*—It was under this shape that Venus was worshipped in Phœnicia. Mr. HIGGINS thinks it was the *Venus Urania*, or Celestial Venus; in allusion to which, the Phœnician grocers first introduced the practice of preserving sugar loaves in blue or sky-coloured paper—he also believes that the *conical* form of the original grenadiers' caps was typical of the loves of Mars and Venus.

[32] Ver. 107. *Parabola*—The curve described by projectiles of all sorts, as bombs, shuttlecocks, &c.

Still e'er she reach it from his polish'd side,
Her trembling hands in devious *Tangents* glide.

Not thus HYPERBOLA: —with subtlest art [33]   115
The blue-eyed wanton plays her changeful part;
Quick as her *conjugated axes* move
Through every posture of luxurious love,
Her sportive limbs with easiest grace expand;
Her charms unveil'd provoke the lover's hand: —   120
Unveil'd, except in many a filmy ray
Where light *Asymptotes* o'er her bosom play,[34]
Nor touch her glowing skin, nor intercept the day.

Yet why, ELLIPSIS, at thy fate repine?[35]
More lasting bliss, securer joys are thine.   125
Though to each fair his treacherous wish may stray,
Though each in turn, may seize a transient sway,
'Tis thine with mild coercion to restrain,
Twine round his struggling heart, and bind with endless chain.

Thus, happy France! in thy regenerate land,   130
Where TASTE with RAPINE saunters hand in hand;
Where nursed in seats of innocence and bliss,
REFORM greets TERROR with fraternal kiss;
Where mild PHILOSOPHY first taught to scan
The *wrongs* of PROVIDENCE, and *rights* of MAN;   135
Where MEMORY broods o'er FREEDOM's earlier scene,
The *Lantern* bright, and brighter *Guillotine*; —
Three gentle swains evolve their longing arms,
And woo the young REPUBLIC's virgin charms:
And though proud *Barras* with the fair succeed,   140
Though not in vain the Attorney *Rewbell* plead,
Oft doth the impartial nymph their love forego,
To clasp thy crooked shoulders, blest *Lepaux*!

[33] Ver. 115. *Hyperbola*—Not figuratively speaking, as in rhetoric, but mathematically; and therefore blue-eyed.

[34] Ver. 122. *Asymptotes*— 'Lines which, though they may approach still nearer together, till they are nearer than the least assignable distance, yet being still produced infinitely, will never meet." —Johnson's Dictionary.

[35] Ver. 124. *Ellipsis*—A curve, the revolution of which on its axis produces an Ellipsoid, or solid, resembling the eggs of birds, particularly those of the gallinaceous tribe. *Ellipsis* is the only curve that embraces the Cone.

So, with dark dirge athwart the blasted heath,
Three Sister Witches hail'd the appall'd Macbeth.   145

So the *Three* Fates beneath grim Pluto's roof,
Strain the dun warp, and weave the murky woof;
'Till deadly Atropos[xxvi] with fatal sheers
Slits the thin promise of the expected years,
While 'midst the dungeon's gloom or battle's din,   150
Ambition's victims perish as they spin.
Thus, the *Three* Graces on the Idalian green,
Bow with deft homage to Cythera's Queen;[xxvii]
Her polish'd arms with pearly bracelets deck,
Part her light locks, and bare her ivory neck;   155
Round her fair form etherial odours throw,
And teach the unconscious zephyrs where to blow;
Floats the thin gauze, and glittering as they play,
The bright folds flutter in phlogistic[xxviii] day.

So, with his Daughters *Three,* the unscepter'd Lear   160
Heaved the loud sigh, and pour'd the glistering tear;
His Daughters *Three,* save one alone, conspire
(Rich in his gifts) to spurn their generous Sire;
Bid the rude storm his hoary tresses drench,
Stint the spare meal, the Hundred Knights retrench;[xxix]   165
Mock his mad sorrow, and with alter'd mien
Renounce the daughter, and assert the queen;
A father's griefs his feeble frame convulse,
Rack his white head, and fire his feverous pulse;
Till kind Cordelia soothes his soul to rest,   170
And folds the Parent-Monarch to her breast.

Thus some fair Spinster grieves in wild affright,
Vex'd with dull megrim, or vertigo light;
Pleased round the fair *Three* dawdling doctors stand,
Wave the white wig, and stretch the asking hand,   175
State the grave doubt, —the nauseous draught decree,
And all receive, though none deserve, a fee.

So down thy hill, romantic Ashbourn[xxx], glides
The Derby dilly[xxxi], carrying *Three* INSIDES.
One in each corner sits, and lolls at ease;   180

With folded arms, propt back, and outstretch'd knees;
While the press'd *Bodkin,* punch'd and squeez'd to death,
Sweats in the midmost place, and scolds, and pants for breath.

[*To be continued.*]

No. XXVI.

May 7.

LOVES OF THE TRIANGLES

The frequent solicitations which we have received for a continuation of the Loves of the Triangles, have induced us to lay before the Public (with Mr. Higgins's permission) the concluding lines of the Canto. The catastrophe of Mr. and Mrs. Gingham, and the Episode of Hippona, contained, in our apprehension, several reflections of too free a nature. The Conspiracy of Parameter and Abscissa against the Ordinate, is written in a strain of Poetry so very splendid and dazzling, as not to suit the more tranquil majesty of diction which our Readers admire in Mr. Higgins. We have therefore begun our Extract with the Loves of the Giant Isosceles, and the Picture of the Asses' Bridge, and its several Illustrations.

CANTO I.

EXTRACT.

'Twas thine alone, O youth of giant frame,
Isosceles![36] that rebel heart to tame!   185
In vain coy Mathesis[37] thy presence flies:[xxxii]

[36] *Isosceles*—An equi-crural Triangle. It is represented as a *Giant,* because Mr. HIGGINS says he has observed that procerity is much promoted by the equal length of the legs, more especially when they are long legs.
[37] *Mathesis*—The doctrine of Mathematics—Pope calls her *mad Mathesis.*—Vide *Johnson's Dictionary.*

Still turn her fond hallucinating[38] eyes;
Thrills with *Galvanic* fires[39] each tortuous nerve,[xxxiii]
Throb her blue veins, and dies her cold reserve;
—Yet strives the fair, till in the Giant's breast   190
She sees the mutual passion flame confess'd:
Where'er he moves she sees his tall limbs trace
Internal Angles[40] *equal at the Base*;
Again she doubts him: but *produced at will*,
She sees *the external Angles equal still.*   195

Say, blest Isosceles! what favouring pow'r,
Or love, or chance, at night's auspicious hour,
While to the *Asses'-Bridge*[41] entranced you stray'd,
Led to the *Asses'-Bridge* the enamour'd maid?
—The *Asses'-Bridge,* for ages doom'd to hear   200
The deafening surge assault his wooden ear,
With joy repeats sweet sounds of mutual bliss,
The soft susurrant sigh, and gently-murmuring kiss.

So thy dark arches, *London Bridge,* bestride
Indignant Thames, and part his angry tide,   205
There oft-returning from those green retreats,
Where fair *Vauxhallia* decks her sylvan seats; —[xxxiv]
Where each spruce nymph, from city compters[xxxv] free,

---

[38] *Hallucinating*—The disorder with which Mathesis is affected, is a disease of *increased volition,* called *erotomania,* or *sentimental love.* It is the fourth species of the second genus of the first order and third class; in consequence of which Mr. Hackman shot Miss Ray in the lobby of the playhouse. —Vide *Zoonomia,* Vol. II. p. 363, 365.

[39] *Galvanic Fires*—Dr. Galvani is a celebrated philosopher at Turin. He has proved that the electric fluid is the proximate cause of nervous sensibility; and Mr. HIGGINS is of opinion, that by means of this discovery, the sphere of our disagreeable sensations may be, in future, considerably enlarged. "Since dead frogs (says he) are awakened by this fluid, to such a degree of posthumous sensibility, as to jump out of the glass in which they are placed, why not men who are sometimes so much more sensible when alive? And if so, why not employ this new stimulus to deter mankind from dying (which they so pertinaciously continue to do) of various old-fashioned diseases, notwithstanding all the brilliant discoveries of modern philosophy, and the example of Count Cagliostro?"

[40] *Internal Angles,* &c. —This is an exact versification of Euclid's 5th theorem. —Vide *Euclid in loco.*

[41] *Asses' Bridge*—Pons Asinorum—The name usually given to the before-mentioned theorem—though, as Mr. HIGGINS thinks, absurdly. He says, that having frequently watched companies of asses during their passage of a bridge, he never discovered in them any symptoms of geometrical instinct upon the occasion. —But he thinks that with Spanish asses, which are much larger, (vide *Townsend's Travels through Spain),* the case may possibly be different.

Sips the froth'd syllabub or fragrant tea;
While with sliced ham, scrap'd beef, and burnt champagne,   210
Her 'prentice lover soothes his amorous pain;
—There oft, in well-trimm'd wherry, glide along
Smart beaux and giggling belles, a glittering throng;
Smells the tarr'd rope—with undulation fine
Flaps the loose sail—the silken awnings shine;   215
"Shoot we the bridge!"—the venturous boatmen cry—
"Shoot we the bridge!"—the exulting fare[42] reply.
—Down the steep fall the headlong waters go,
Curls the white foam, the breakers roar below.
—The veering helm the dextrous steersman stops,   220
Shifts the thin oar, the fluttering canvas drops;
Then with closed eyes, clench'd hands, and quick-drawn breath,
Darts at the central arch, nor heeds the gulf beneath.
—Full 'gainst the pier the unsteady timbers knock,
The loose planks starting own the impetuous shock;   225
The shifted oar, dropt sail, and steadied helm,
With angry surge the closing waters whelm—
—Laughs the glad Thames, and clasps each fair one's charms
That screams and scrambles in his oozy arms.   230
—Drench'd each smart garb, and clogg'd each struggling limb,
Far o'er the stream the Cockneys sink or swim;
While each badged boatman[43] clinging to his oar,
Bounds o'er the buoyant wave, and climbs the applauding shore. [xxxvi]

So, towering Alp! from thy majestic ridge[44]   235
Young Freedom gazed on Lodi's blood-stained *Bridge*;[xxxvii]
—Saw, in thick throngs, conflicting armies rush,
Ranks close on ranks, and squadrons squadrons crush;
—Burst in bright radiance through the battle's storm,
Waved her broad hands, display'd her awful form;   240
Bade at her feet regenerate nations bow,

[42] *Fare*—A person, or a number of persons, conveyed in a hired vehicle by land or water.
[43] *Badged boatman*—Boatmen sometimes wear a *badge,* to distinguish them: especially those who belong to the Watermen's Company.
[44] *Alp or Alps*—A ridge of mountains which separate the North of Italy from the South of Germany. They are evidently primeval and volcanic, consisting of granite, toadstone, and basalt, and several other substances containing animal and vegetable recrements, and affording numberless undoubted proofs of the infinite antiquity of the earth, and of the consequent falsehood of the Mosaic Chronology.

And twined the wreath round Buonaparte's brow.
—Quick with new lights, fresh hopes, and alter'd zeal,
The slaves of Despots dropt the blunted steel:
Exulting Victory own'd her favourite child,   245
And freed Liguria[xxxviii] clapt her hands and smiled.

   Nor long the time ere Britain's shores shall greet
The warrior-sage with gratulation sweet:
Eager to grasp the wreath of Naval Fame,
The GREAT REPUBLIC plans the *Floating Frame!*   250
—O'er the huge plane gigantic *Terror* stalks,
And counts with joy the close-compacted balks:
Of young-eyed *Massacres* the Cherub crew,
Round their grim chief the mimic task pursue;
Turn the stiff screw,[45] apply the strengthening clamp,   255
Drive the long bolt, or fix the stubborn cramp,
Lash the reluctant beam, the cable splice,
Join the firm dove-tail with adjustment nice,
Through yawning fissures urge the willing wedge,
Or give the smoothing adze a sharper edge.   260
—Or group'd in fairy bands with playful care,
The unconscious bullet in the furnace bear;
Or gaily tittering, tip the match with fire,
Prime the big mortar, bid the shell aspire;
Applaud with tiny hands and laughing eyes,   265
And watch the bright destruction as it flies.

   Now the fierce forges gleam with angry glare—
The windmill waves[46] his woven wings in air;
Swells the proud sail, the exulting streamers fly,
Their nimble fins unnumber'd paddles ply: 270
—Ye soft airs breath, ye gentle billows waft,

---

[45] *Turn the stiff screw, &c.* —The harmony and imagery of these lines are imperfectly imitated from the following exquisite passage in *The Economy of Vegetation*:

    Gnomes, as you now dissect, with hammers fine,
    The granite rock, the nodul'd flint calcine;
    Grind with strong arm, the circling Chertz betwixt,
    Your pure ka—o—lins and Pe—tunt—ses mixt.

                                Canto 2, l. 297

[46] *The Windmill.* &c. —This line affords a striking instance of the sound conveying an echo to the sense. —I would defy the most unfeeling reader to repeat it over without accompanying it by some corresponding gesture imitative of the action described. —*Editor.*

And fraught with Freedom, bear the expected Raft!
—Perch'd on her back, behold the Patriot train,
Muir, Ashley, Barlow, Tone, O'Connor, Paine;^{xxxix}
While Tandy's hand directs the blood-empurpled rein.^{xl}   275

Ye Imps of Murder, guard her angel form,
Check the rude surge, and chase the hovering storm;
Shield from contusive rocks her timber limbs,
And guide the sweet Enthusiast[47] as she swims!

—And now, with web-foot oars, she gains the land,   280
And foreign footsteps press the yielding sand:
—The Communes spread, the gay Departments smile,
Fair Freedom's Plant o'ershades the laughing isle:
—Fired with new hopes the exulting peasant sees
The Gallic streamer woo the British breeze;   285
While, pleased to watch its undulating charms,
The smiling infant[48] ^{xli}spreads his little arms.

Ye Sylphs of DEATH, on demon pinions flit
Where the tall Guillotine is raised for Pitt:
To the poised plank tie fast the monster's back,[49]   290
Close the nice slider, ope the expectant sack;
Then twitch, with fairy hands, the frolic pin—
Down falls the impatient axe with deafening din;
The liberated head rolls off below,[50] ^{xlii}
And simpering Freedom hails the happy blow!295

---

[47] *Sweet Enthusiast,* &c. —A term usually applied in allegoric or technical poetry, to any person or object to which no other qualifications can be assigned. —*Chambers's Dictionary.*

[48] *The smiling infant*—Infancy is particularly interested in the diffusion of the new principles. —See the "Bloody Buoy" —see also the following description and prediction:

> Here Time's huge fingers grasp his giant mace,
> And dash proud Superstition from her base;
> Rend her strong towers and gorgeous fanes, &c.
> &c. &c. &e. &c.
> While each light moment, as it passes by,
> With feathery foot and pleasure-twinkling eye,
> Feeds from its baby-hand with many a kiss
> The callow nestlings of domestic bliss.

Botanic Garden

[49] *The monster's back*—Le Monstre Pitt, l'Ennemi du Genre humain. —See Debates of the Legislators of the Great Nation *passim.*

[50] Atque illud prono præceps agitur decursus. —*Catullus.*

Notes

i Mr Higgins is William Godwin.

ii An earlier satire on Richard Payne Knight.

iii Pope, *Essay on Man*, p. 249 (I, 293–4): 'And, spite of Pride, in erring Reason's spite, / One truth is clear, "Whatever is, is RIGHT." '

iv Godwin took the concept of perfectibility from Jean-Jacques Rousseau.

v 'Oxygene' was the recent French term for the gas known in Britain as 'dephlogisticated air'.

vi In *The Loves of the Plants*, Darwin promised to 'inlist Imagination under the banner of Science.'

vii The binomial system of classification introduced by Carl Linnaeus.

viii Asses' Bridge (*Pons Asinorum* in Latin) refers to Euclid's fifth proposition in Book I of his *Elements of Geometry*, which defines the properties of the angles of an isosceles triangle. As the first proposition requiring mathematical intelligence, it provides a bridge to the more difficult ideas that follow.

ix Isaac Newton's version of calculus was called fluxions.

x This may refer to the trochee, a metrical foot in poetry consisting of a stressed syllable followed by an unstressed one (trochiaic tetrameter refers to lines of four such feet).

xi French Revolutionaries: Barras was perceived as the most dangerous.

xii Public stage-coach.

xiii Gingham is an important character in Frederick Reynolds' play *The Rage* (1794), a successful romantic comedy performed at Covent Garden, which satirizes obsession with fashion, and involves an accident with a carriage and runaway horses. Reynolds' *How to grow rich* was printed at the Anti-Jacobin Press, which was probably associated with the *Anti-Jacobin Review*.

xiv Digamma: discontinued sixth letter of Greek alphabet. Pope, p. 561 (*The Dunciad*, IV, 217–18): 'While tow'ring o'er your Alphabet, like Saul, / Stands our Digamma, and o'ertops them all.'

xv Pictures of Hippona, the Greek goddess who protected horses, were often hung in stables.

xvi See note viii.

xvii Battle of Lodi, 10 May 1796, between Napoleon and retreating Austrian troops, which did much to establish Napoleon's reputation as a brilliant commander.

xviii Osculation was originally a seventeenth-century word meaning a kiss.

xix Hyde Park.

xx Authors' footnote quotes Horace, *Odes* 1.9.22: 'The winsome smile of a girl from a corner'.

xxi See note x.

xxii A pun based on a schoolboy howler in translating Ovid. In the authors' note, Howel Dda was a tenth-century chieftain who codified Welsh law.

xxiii Frere wrote most of this instalment, but the last three lines were by Canning.

xxiv Menecmus was a mathematician of the 4th century BC who studied conic sections before Euclid.

xxv Caesar erected a pillar for Pompey on the Egyptian seashore where he had been murdered.

xxvi One of the three fates, who cut the thread of life; Milton, *Lycidas* ll. 75–6: 'Comes the blind *Fury* with th'abhorred shears, / And slits the thin-spun life'.

xxvii The three Graces were daughters of Zeus. Paris was abandoned on the forested Mount Ida, where he awarded the prize of beauty to Aphrodite, queen of Cythera.

xxviii See note v.

xxix The King of the Hundred Knights is a character from the Arthurian legends

xxx Small town in Staffordshire. Brooke Boothby had lived at Ashbourne Hall, but it was then the home of one of Canning's relatives, whom he often visited. Darwin's daughters' school was at Ashbourne.

xxxi The Derby Dill was a nineteenth-century political group. *Western Mail*, 25 Oct 1869, makes the link in an obituary for the Earl of Derby.

xxxii Mathesis: hypothetical universal mathematical science proposed by Leibniz and Descartes. Johnson's *Dictionary* quotes from Pope's *Dunciad*, IV, 31–2: 'Mad *Mathesis* alone was unconfin'd / Too mad for mere material chain to bind.'

xxxiii Luigi Galvani's experiments demonstrated the existence of animal electricity, but this was before Alessandro Volta introduced current electricity.

xxxiv Vauxhall Gardens.

xxxv Debtors' prisons.

xxxvi Five people had been killed in a sailing accident on the Thames the previous week.

xxxvii See note xvii.

xxxviii Region of NW Italy (capital Genoa).

xxxix Thomas Muir, Arthur O'Connor, John Ashley, Joel Barlow, Wolfe Tone, Thomas Paine: prominent sympathizers of the French Revolution.

xl James Tandy, the hero of 'The Wearing of the Green': an Irish revolutionary who fled to France and landed in Donegal.

xli William Cobbett emphasized revolutionary atrocities with examples of children seeing their fathers executed or playing with toy guillotines.

xlii 'See, on and down it swiftly rolls and runs': Catullus, *Poems*, 65.23.

NOTES

Introduction

1 Nabokov, p. 209.

2 Traweek.

3 *Times Literary Supplement*, 18 June 2004, pp. 4–5 (review in 2004 by Andrew Scull of Robert Merton and Elinor Barber, *The Travels and Adventures of Serendipity*, Princeton University Press).

4 Southam, pp. 15–16.

5 Wilberforce, pp. 254–5: the debate was in the same year as publication, 1860.

6 On 17 September 2008, I found thirty-one hits on 'loves of the triangles' in '19th century British Library newspapers' (http://find.galegroup.com/bncn) and eight on '19th century UK periodicals' (http://find.galegroup.com/ukpc/retrieve). It was still referred to in the twentieth century (eg. *Illustrated London News*, 16 November 1907, p. 725; reviews by G. K. Chesterton).

7 *Daily News*, 22 November 1888.

8 Long after painstakingly transcribing the poem, I fortuitously found a modern annotated edition (in Stones, a five-volume anthology): unfortunately, standard catalogue searches had failed to find it.

9 Quoted King-Hele, *Erasmus Darwin and the Romantic Poets*, p. 26. In Walpole's opinion, Darwin's poem was rivalled only by the *Arabian Nights* and King's College chapel in Cambridge.

10 King-Hele, *The Collected Letters of Erasmus Darwin*, p. 358 (letter to James Watt of 19 January 1790).

11 Richard Lovell Edgeworth quoted Charles Darwin, p. 67.

12 I might as well confess now that, unlike Walpole, I find Darwin's poetry tedious. I am certainly not alone in this: his literary style was old-fashioned even at the time, and my personal dislike reflects the taste of my own contemporaries. Unsurprisingly, my enthusiasm during this project has flagged occasionally, although—like almost anything that you study closely—Darwin's imagery became more intriguing as I unravelled some of its complexities. If you would like to see Darwin's poems, they are available online and in cheap paperback reproductions. There is no substitute for reading things for yourself, and you may well find that your interpretations differ substantially from mine. In any

case, I make no claim to a systematic exposition; instead, I have concentrated on those sections that are especially relevant for understanding 'The Loves of the Triangles'.

13 *Glasgow Herald*, 18 July 1899.

14 *Punch*, 25 September 1869.

15 *Liverpool Mercury*, 17 August 1878; *Pall Mall Gazette*, 31 January 1870 (I have restored the capitalization of the original).

16 Darnton, pp. 74–104.

The Loves of the Triangles

1 Stevens, p. 69 (20 February 1793).

2 Stevens, p. 33 (20 June 1792).

3 Letter to Josiah Wade of 27 January 1796, quoted King-Hele, *Darwin: A Life of Unequalled Achievement*, p. 301.

4 King-Hele, *Darwin: A Life*, p. 277.

5 Quoted Uglow, *The Lunar Men*, p. 40, from Boswell's biography of Samuel Johnson.

6 Darwin, *The Economy of Vegetation*, p. 116 (III, 20, footnote).

7 Salinger, p. 1.

8 Pearson, pp. 253–4 (letter from Anna Seward to William Hayley, 7 March 1803).

9 Charles Darwin, p. 35 (James Keir).

10 Darwin, *The Loves of the Plants*, pp. 129–30 (interlude III).

11 Poem to his sister quoted King-Hele, *Darwin: A Life*, p. 8. The most detailed biography yet written, this invaluable book is based on extremely thorough research; I have relied heavily on it throughout, although I have given page references only for quotations and unusual details. My other two major biographical sources are McNeil, *Under the Banner*, and Uglow, *Lunar Men*. Studies of Darwin's poetry include: Browne, 'Botany for Gentlemen'; Danchin, 'Erasmus Darwin's Scientific and Poetic Purpose in The Botanic Garden'; and McNeil, 'Scientific Muse'.

12 Brewer, pp. 573–612; Wilson, 'Collecting the Instruments of Life around Me'.

13 Seward, pp. 2–3.

14 Uglow, *Lunar Men*, provides a marvellous account.

15 Stevens, p. 128 (19 January 1794).

16 King-Hele, *Darwin: A Life*, p. 161.

17 King-Hele, *Collected Letters*, p. 181 (letter to James Watt, 6 January 1781).

18 First and last stanzas of 'Hunting Song': Darwin, *To Elizabeth, with Love*, p. 22. I have changed 'discontent' to 'disconcert', which enables a rhyme with 'virtue'.

19 Darwin, *Loves of the Plants*, p. 1 (I, 1–4). For his influence, see King-Hele, *Erasmus Darwin and the Romantic Poets*; for literary studies of his poetry, see Hassler, Logan. For a useful brief account of his later books, see Priestman, 'Introduction'.

20 Porter, 'Erasmus Darwin'.

21 Uglow, 'But What About the Women?'.

22 Darwin, *Phytologia*, pp. 556–60 (quotations pp. 560 and 556).

23 Hochschild, p. 237; Grenby, especially pp. 1–12.

24 Quoted from *Letters on a Regicide Peace* (1796) on p. 9 of Burke, *Reflections on the Revolution in France* (introduction by Conor Cruise O'Brien).

25 Most of my information on the editions and the contributors is taken from St Clair, pp. 192–6, Stones, pp. xlv–lx, and De Montluzin, pp. 21–5.

26 Pope, p. 250 (*An Essay on Man*, Ep. II, 1–2).

27 Darwin, *Loves of the Plants*, pp. 10 and 12 (I, 85–6, 97–8).

28 Grafton, pp. 111–21.

29 'Loves of the Triangles', 53–7: see Appendix. This became a favourite extract for nineteenth-century commentators. *Coctilibus muris* means 'walls of baked mud-bricks', but the authors deliberately mistranslate *muris* as 'mice' (*mus*, *muris*).

30 St Clair, pp. 205 and 223; Bewell, '"Jacobin Plants"'.

31 *Encycloaedia Britannica* (1801), vol. 1, p. iv. See Morrell.

32 Stevens, p. 399 (9 November 1796).

33 Burke (quotations pp. 287 and 286). Buck, Crosland.

34 Darwin, *Loves of the Plants*, p. 8 (I, 71–2).

35 Stevens, p. 128 (19 January 1794).

36 Darwin, *Loves of the Plants*, II, 1–2, 87–8.

37 *Analytical Review* 15 (1793), p. 289 (William Cowper). See Robinson, 'Erasmus Darwin's *Botanic Garden*'.

38 Quoted King-Hele, *Darwin: A Life*, p. 265.

39 Letter from William Godwin to Percy Shelley, 10 December 1812, quoted Marshall, p. 303.

40 Quoted King-Hele, *Darwin: A Life*, p. 265.

41 From *English Bards and Scotch Reviewers*, quoted Logan, p. 93.

42 Moody, pp. 10–11 (Elizabeth Moody, née Greenly: see Lonsdale, pp. 401–7).

43 Wordsworth; King-Hele, *Erasmus Darwin and the Romantic Poets*, pp. 62–87.

44 Quoted King-Hele, *Erasmus Darwin and the Romantic Poets*, p. 71 (*Edinburgh Magazine* 2 [1818], p. 313).

45 Grafton, *Footnote*, pp. 69–70.

46 Moody, pp. 9–10 (Elizabeth Moody, née Greenly: see Lonsdale, pp. 401–7).

47 Falconer, p. 10 (Proem, 233–8).

48 Priestman, *Romantic Atheism*, pp. 44–79, to which I am indebted.

49 This account is closely based on the entry in the *New Oxford Dictionary of National Biography* (*NDNB*) by Michael Franklin. My other major sources are the works by Cannon and Franklin listed in the bibliography.

50 Quoted Franklin (from *Critical Review* 59 [1785], pp. 19–21).

51 Cannon, *Life and Mind*, p. 336.

52 Letter from Matthew Guthrie to Joseph Banks of June 1801: Banks, vol. 5, p. 96 (letter 1595).

53 The biographical aspects of this account are closely based on the entry in the *New Oxford Dictionary of National Biography* by C. Stumpf-Condrey and S. J. Skedd. My other major sources are Ballantyne and Messman.

54 Quoted Hussey, p. 106.

55 Quoted Klingender, p. 77 (from *Annals of Agriculture*, 1785).

56 Smith, pp. 182–5.

57 Letter to Banks from Johan Blumenbach: Banks, vol. 4, p. 264.

58 Quoted Fabricant, p. 126.

59 McNeil, *Under the Banner of Science*, pp. 32–9; Hussey; Darwin, *Loves of the Plants*, pp. viii–ix (Proem).

60 Franklin (*NDNB* entry on Jones).

61 *Analytical Review* 15 (1793), p. 289.

62 Cannon, 'Sir William Jones, Sir Joseph Banks, and the Royal Society' (quotation p. 217); Fulford, pp. 120–2.

63 Elliott, pp. 1–11.

64 Letter of 6 July 1816 to Thomas Knight: Banks, vol. 6, p. 208.

65 Letter of 23 January 1816 to Thomas Knight: Banks, vol. 6, p. 197.

66 Quoted McNeil, *Under the Banner*, p. 98.

67 Fara and Money, pp. 560–5; Hopkins; Johnson and Wilson; Priestman, 'Lucretius in Romantic and Victorian Britain'.

68 Quoted Hopkins, p. 257 (from Creech's preface).

69 Quoted Johnson and Wilson, p. 141.

70 Good, vol. 1, pp. 9–10 (notes) and *passim*.

71 The following discussion leans heavily on Priestman, *Romantic Atheism*, pp. 44–79.

72 Jones, p. 69 (Introduction to 'A Hymn to Lacshmi', published 1789).

73 Quoted Priestman, *Romantic Atheism*, p. 53.

74 Darwin, *Temple of Nature*, p. 4 (I, 15–20).

75 Quoted Priestman, *Romantic Atheism*, p. 44.

76 Jones, p. 79 ('The Hymn to Bhavani,' from stanzas 1 and 3).

77 Mathias, part I, pp. 12–13. In the *Anti-Jacobin* 'New Morality', Mathias was hailed as 'the nameless bard'.

78 Quoted Priestman, *Romantic Atheism*, p. 79.

The Loves of the Plants

1 Johnson, vol. 3, pp. 271–6; Harvey.

2 Matthews, 'The Surprising Success of Dr. Armstrong'. In 1795 it was published with *The Triumphs of Temper*, a successful poem for women by Darwin's friend William Hayley.

3 Fabricant (Horace Walpole quoted p. 110).

4 Pope, p. 316 (*Epistle to Burlington*, 51–4).

5 King-Hele, *Darwin: A Life*, pp. 134–6, 149–51.

6 Darwin, *Economy of Vegetation*, p. 3 (I, 26 note). Matthews, 'Impurity of Diction'.

7 Darwin, *To Elizabeth*, p. 27 (from 'Platonic Epistle to a Married Lady').

8 Seaton, 'Towards a Historical Semiotics of Literary Flower Personification'.

9 Quoted Lloyd, p. 428 (*Morning Post*, 7 July 1785, p. 2).

10 Quoted p. 133 of Bewell, '"Jacobin Plants"'.

11 Shteir, *Cultivating Women* (quotations pp. 37 and 9).

12 Quoted King-Hele, *Darwin: A Life*, p. 150.

13 Quoted King-Hele, *Darwin: A Life*, p. 151.

14 Jackson, W., p. 286.

15 King-Hele, *Collected Letters*, p. 184 (letter to Thomas Day of 16 May 1781).

16 These numbers were correct on 18 August 2011, but continually change as new books are acquired.

17 Bradbury, pp. vii–xv, 316–18.

18 Seward, p. 134.

19 Seward, pp. 138 and 143. Seward's dates differ by a few weeks from those in other copies of these letters: see King-Hele, *Collected Letters*, p. 179.

20 Vickery, p. 228.

21 Quoted Browne, 'Botany for Gentlemen', p. 595, n. 6.

22 Snow, p. 9.

23 Powell, pp. 58–60 (see Darwin's third canto).

24 Polwhele, *The Unsex'd Females*, p. 21.

25 Claudian, *Epithalamium for Emperor Honorius*, 65–9. I am grateful to Helen van Noorden for this loose translation: 'The leaves live for love; in every deep grove, the happy tree is in love: the palms bend down to mate with each other,

poplar sighs with passion for poplar, and plane tree whispers to plane, and alder to alder.'.

26 'Loves of the Triangles', Introduction (see Appendix).

27 Darwin, *Loves of the Plants*, p. vii (Proem).

28 Quoted Fabricant, p. 127 (from a letter to Edward Blount, June 1725).

29 Jackson, H., pp. 8–18.

30 Shteir, 'Iconographies of Flora'.

31 Bewell, '"On the Banks of the South Sea"', p. 190.

32 Fabricant. Gilbert West wrote the Stowe estate poem.

33 Darwin, *Loves of the Plants*, pp. 16–17 (I, 151–6).

34 See Appendix.

35 Moody, p. 12.

36 Darwin, *Loves of the Plants*, p. 60 (II, 77–84).

37 Darwin, *Loves of the Plants*, p. 62 (II, 101–4).

38 Darwin, *Loves of the Plants*, p. 61 (II, 87 note).

39 'Loves of the Triangles', 53–7: see Appendix. This became a favourite extract for nineteenth-century commentators.

40 Darwin, *Loves of the Plants*, pp. 126–7 (III, 441–4).

41 Darwin, *Loves of the Plants*, p. 178 (IV, 471–2).

42 Ovid, pp. 233–45 (Book X).

43 McNeil, 'The Scientific Muse', pp. 185–93, seems to be the only substantial commentary on Darwin's prose interludes.

44 Darwin, *Love of the Plants*, p. 48 (Interlude I).

45 Jay, pp. 1–49.

46 Hankins and Silverman, pp. 72–112.

47 Darwin, *Loves of the Plants*, p. 127 (III, 455–6).

48 Jones, *Jane Austen*, p. 3.

49 Quoted Vickery, p. 197. The best account of dependent women's lives is Vickery, pp. 184–206.

50 Wollstonecraft, p. 88.

51 Quoted Schiebinger, pp. 22–3.

52 Darwin, *Loves of the Plants*, pp. 4–5 (I, 51–6). My account relies heavily on Browne, 'Botany for Gentlemen'.

53 Darwin, *Loves of the Plants*, p. 4 (I, 45–50) (*Callitriche*, stargrass).

54 Plumptre, p. 24. See George, especially pp. 117–19.

55 Plumptre, p. xii.

56 *Encyclopaedia Britannica*, quoted Thornton, p. 9.

57 Binhammer.

58 Letter from Darwin of 17 July 1792, in Polwhele, *Traditions and Recollections*, p. 299.

59 Quoted by Gina Luria in Polwhele, *Unsex'd Females*, 'Introduction', p. 9.

60 Polwhele, *Unsex'd Females*, p. 8 (note).

61 Polwhele, *Unsex'd Females*, pp. 25–6.

62 Wollstonecraft, p. 79.

63 Binhammer, p. 409.

64 Polwhele, *Unsex'd Females*, pp. 8–10.

65 Mathias, part I, pp. 12–13.

66 Seaton, 'Historical Semiotics'.

67 *Monthly Review* 10 (1793), p. 183. In the 1791 edition of De la Croix, the translator draws attention to similarities with Darwin; the original of 1728 drew on Sebastien Vaillant rather than Linnaeus.

68 Crabbe, p. 122 (ll. 175–8).

69 Darwin, *Loves of the Plants*, p. 32 (I, 309–12).

70 Chamberlain, p. 849 (from 'The Parish Register').

71 Chamberlain; see also Garfinkle.

72 Seaton, *The Language of Flowers*, pp. 36–79.

73 Jones, 'Enchanted Fruit', p. 42; see Fulford. The poem was published in the first volume of *Asiatick Miscellany*.

74 Maria Jacson sent Darwin her botanical poems to assess: King-Hele, *Collected Letters*, pp. 482–3. See also King-Hele, *Romantic Poets*. Although Anon (*The Golden Age*) is attributed to Darwin, it is clearly a satirical attack (a footnoted poem in heroic couplets) on both Darwin and Thomas Beddoes, as indicated by the D—n on the title page. In the Cambridge University Library copy, someone has written 'Richard Saunders, D D' in black ink at the end. Saunders was indeed a clergyman associated with Oxford (where the poem was published) in the 1790s, but he seems to have been a sober scholar of law and oriental languages.

75 Rowden, p. 27. See also Montolieu and Moody, pp. 8–12 ('To Dr Darwin, on Reading his Loves of the Plants'; Elizabeth Moody, née Greenly: see Lonsdale, pp. 401–7).

76 Knight, p. 55 (III, 150–3).

77 Polwhele, *Traditions*, pp. 535–6; Mathias, part 1, ll. 133–6 (pp. 20–1).

78 Mathias, part 1, ll. 133–6 (pp. 20–1).

79 Jenkins and Sloan, pp. 216–18.

80 Letter of 17 July 1781: Banks, *Correspondence*, vol. 1, p. 276.

81 Jenkins and Sloan, pp. 238–9. I am grateful to Eamon Duffy for telling me about a sixteenth-century parallel: the phallic shrine of St Vallery in Picardy, whose wax reproductions are discussed by Thomas More in *Dialogue concerning Heresies*.

82 Mitter, pp. 84–6.

83 My main source is Redford, pp. 113–28.

84 Priestman, *Romantic Atheism*, pp. 56–8.

85 Cairncross, pp. 157–65; Yeazell, pp. 99–101. *Thelyphthora* appeared as two volumes in 1780, and as three volumes in the longer edition of 1781.

86 Matthew 25:1–13.

87 Matthews, 'Impurity of Diction', pp. 67–73.

88 Andrew, pp. 411–15 (Genesis 5:19).

89 Wollstonecraft, p. 164.

90 Cowper, pp. 317–21 (quotation p. 318, 'Anti-Thelyphthora', ll. 58–61).

91 Darwin, *Poems of Lichfield and Derby*, p. 35.

92 Lloyd, quotation p. 421.

93 Lloyd, p. 436.

94 Rice-Oxley, p. 102 ('Progress of Man', Canto XXIII); see Knight, p. 56 (III, 170–1): 'When, if changed tempers made affection cease,/The power that join'd them could again release.'.

95 Bindman, *From Ape to Apollo*, and Nussbaum are my major sources for this discussion.

96 Wollstonecraft, p. 83.

97 Horace Walpole, quoted Harman, p. 209.

98 Adam Ferguson, quoted Nussbaum, p. 8.

99 Godfrey, p. 218; Wood, pp. 155–6.

100 Quoted Ernst, p. 11.

101 Quoted Painter, pp. 70–1 (talk in 1795, publication 1799).

102 Hume, p. 213 (from 'On National Characters').

103 Painter, pp. 43–58.

104 Horace Walpole, quoted Harman, p. 209.

105 Madan, I, p. 75n.

106 Nussbaum, pp. 73–94; Aldridge; Cairncross, pp. 157–65.

107 Nussbaum, p. 226, n. 15.

108 Hume, pp. 180–95 ('Of Polygamy and Divorces').

109 Matthews, 'Impurity', pp. 67–73.

110 Brown, p. 337.

111 Shyllon, pp. 105–8. Darwin, *Zoonomia*, I, 514 and I, 569 (addition).

112 Nussbaum, pp. 169–70.

113 Sancho, quoted Nussbaum, p. 75.

114 Montagu, vol. 1, p. 314 (including note 3).

115 Yeazell, p. 29.

116 Montagu, *Letters*, vol. I, 314 ('Letter to Lady ——', 1 April 1717).

117 Jones, quoted Leask, p. 2.

118 Sir Robert Gordon Menzies, 1939 (*Oxford Dictionary of Quotations*).

119 Yohannan.

120 Jones, W., p. 47.

121 George, p. 113.

122 Fulford, pp. 27–30; Darwin, *Loves of the Plants*, p. 32 (I, 309–12).

123 Darwin, *Loves of the Plants*, p. 168 (IV, 335–40).

124 Quotation from Smith, p. 46.

125 Mavor, vol. 1, p. 381.

126 Darwin, *Loves of the Plants*, p. 165 (IV, 309–12).

127 Darwin, *Loves of the Plants*, p. 166 (IV, 321–4). See Seaton, *Language of Flowers*, pp. 36–79, and Yohannan, pp. 146–51.

128 'Loves of the Triangles', 86–104: see Appendix.

The Economy of Vegetation

1 Quoted Leask, p. 1.

2 Herbert, pp. 78–80, 103–8.

3 Holmes, p. 252.

4 Wood, pp. 157–9; Jordanova, pp. 21–33; Alexander (plate 11); Bindman, *Shadow of the Guillotine*.

5 Bindman, *Shadow of the Guillotine*, pp. 126–7 (see also p. 89).

6 Quoted Oldfield, p. 33.

7 Quoted Bindman, *Mind-forg'd Manacles*, p. 14.

8 Judith Drake, quoted Nussbaum, p. 17.

9 Thomas, pp. 465–510 (Johnson quoted p. 476).

10 Quoted Uglow, *Lunar Men*, p. 257.

11 King-Hele, *Collected Letters*, p. 134 (letter to Watt, 29 March 1775).

12 Fox, pp. 222–9.

13 Porter, *Enlightenment*. Priestley, quoted Uglow, *Lunar Men*, p. xiv (from Priestley's *Experiments on the Generation of Air from Water*, 1793). This marvellous book is my major source of information for this section, although I have only referenced quotations.

14 Darwin, *Collected Letters*, p. 150 (letter to Matthew Boulton, 5 April 1778).

15 Quoted Fox, p. 222 (letter to Boulton, 2 January 1765).

16 Quoted Porter, *Enlightenment*, p. 424.

17 Uglow, *Lunar Men*, pp. 181–8 (Seward quoted p. 188); Flavell, pp. 99–101.

18 Day and Bicknell, p. 24.

19 Darwin, *Loves of the Plants*, III, 411–18.

20 Catherine Wright, quoted Uglow, 'What about the Women?', p. 164 (letter to Withering, 30 January 1785).

21 Darwin, *Collected Letters*, p. 339 (letter to Robert Darwin, 19 April 1789).

22 Darwin *Collected Letters*, p. 359 (letter to Watt, 19 January 1790).

23 Darwin, *Collected Letters*, pp. 357–64 (31 December 1789 to 24 April 1790).

24 This approach is taken mainly from McNeil, 'Scientific Muse'.

25 Carter, p. 295.

26 Carter, p. 297. Darwin definitely knew that lead is dangerous, because he objected to Richard Polwhele describing a '*leaden* pound' for pressing cider: Letter from Darwin to Polwhele of 17 July 1792, in Polwhele, *Traditions*, p. 396.

27 McKendrick (quotations pp. 33, 39, 34).

28 Uglow, 'What about the Women?'.

29 Darwin, *Collected Letters*, p. 359 (letter to Watt, 19 January 1790).

30 Quoted Uglow, *Lunar Men*, p. 478 (letter to Dr Edward Ash, 1798).

31 Matthews, 'Armstrong'.

32 Letter from Watt to Darwin, 24 November 1789, quoted King-Hele, *Darwin*, p. 244.

33 *Systema* was first published in 1735, but Linnaeus enlarged it in successive editions, the one of 1758 being the most important. *Economy of Nature* (1749) was a shorter treatise.

34 Koerner.

35 Spary.

36 Darwin, *Economy of Vegetation*, pp. 101–2 (note XXXVII on 'Vegetable Respiration').

37 Darwin, *Economy of Vegetation*, p. 1 (I, 1–2).

38 'Loves of the Triangles', 1–4: see Appendix.

39 Darwin, *Economy of Vegetation*, p. 27 (I, 259–62).

40 Darwin, *Economy of Vegetation*, p. 29 (I, 285–8).

41 Darwin, *Economy of Vegetation*, p. 8 (I, 101–2).

42 Quoted Priestman, *Romantic Atheism*, p. 44.

43 Darwin, *Economy of Vegetation*, p. 87 (II, 307–8).

44 Darwin, *Economy of Vegetation*, p. 66 (II, 72–6).

45 Darwin, *Economy of Vegetation*, p. 91 (II, 363–8).

46 Darwin, *Economy of Vegetation*, p. 91 (II, 371–2).

47 'Loves of the Triangles', 235–6: see Appendix.

48 Darwin, *Economy of Vegetation*, pp. 92–3 (II, 385–9).

49 'Loves of the Triangles', 241–4: see Appendix.

50 'Loves of the Triangles', 247–8, 282–3: see Appendix.

51 Darwin, *Economy of Vegetation*, pp. 106–7 (II, 581–4).

52 Darwin, *Economy of Vegetation*, p. 156 (III, 555–8).

53 Letter of 1 August 1772, quoted King-Hele, *Darwin: A Life*, p. 113.

54 King-Hele, *Darwin: A Life*, p. 48.

55 Darwin, *Economy of Vegetation*, p. 142 (III, 347–52).

56 Darwin, *Economy of Vegetation*, p. 132 (III, 201–4).

57 Darwin, *Economy of Vegetation*, pp. 199–200 (IV, 479–84).

58 Hogg, vol. I, p. 63.

59 Darwin, *Economy of Vegetation*, p. 178 (IV, 197–202).

60 Darwin, *Economy of Vegetation*, p. 165 (IV, 45–6).

61 Darwin, *Economy of Vegetation*, p. 185 (IV, 297–300).

62 *Times Literary Supplement*, 19 and 26 August 2011, p. 15 (by Henry Power).

63 Darwin, *Economy of Vegetation*, p. 191 (IV, 377–80).

64 Darwin, *Economy of Vegetation*, p. 210 (IV, 623–4, 627–8).

65 Manuscript with the 1791 edition of *The Botanic Garden* formerly owned by Sir Richard Colt Hoare (Cambridge University Library, Syn.4.79.13).

66 Darwin, *Economy of Vegetation*, pp. 95–6 (II, 413–18, 421–4).

67 Letter from Darwin to Wedgwood, 22 February 1789, Cambridge University Library, DARWIN 227:1, 110, reproduced in King-Hele, *Collected Letters*, pp. 331–3.

68 Godfrey, p. 218; *NDNB* article by Robert Hole on Horsley; Wood, pp. 155–6.

69 Advertisement in *Aris's Birmingham Gazette*, 11 November 1771, reproduced Torrington et al., p. 43.

70 Sprat, pp. 407–8; Davies, pp. 38–46.

71 Francis Hargrave (in 1772) quoted Drescher, pp. 85–6; see also pp. 455–9; Colley.

72 The extensive literature includes Davis (the classic study, reviewed by Robinson, 'Capitalism'), Thomas, Drescher, Inikori and Hochschild. For circulating credit, see Baucum, especially pp. 35–112. For sugar, see Drayton and Abbott. For black rice, see Carney.

73 Malachai Postlethwayt, quoted Williams, p. 51.

74 Hochschild, p. 117.

75 Flavell, pp. 27–61.

76 Quoted Shyllon, p. 14.

77 Gardner: Fitzwilliam Museum, Cambridge. Chelsea porcelain, *c*.1755. C.P.C. 176. See Ellis.

78 Shyllon, pp. 84–113.

79 Quoted with no reference in Shyllon, p. 31.

80 Brown.

81 Hochschild, pp. 41–8.

82 Quoted Shyllon, p. 34, from *A Dialogue between a Justice of the Peace and a Farmer*, 1785.

83 Quoted Hochschild, pp. 89–90.

84 Farrell et al., pp. 308–13.

85 Letter from Darwin to Wedgwood, 13 April 1789, Cambridge University Library, DARWIN 227:1, 112, reproduced in King-Hele, *Collected Letters*, p. 338.

86 Clarkson, vol. 1, p. 376.

87 Abbott, pp. 187–245, especially p. 239.

88 Darwin, *Phytologia*, p. 77.

89 Wood, pp. 14–40; Baucom, pp. 9–18; Boddice, pp. 25–154.

90 Quoted Flavell p. 133, from *Fragment of an Original Letter* (1776).

91 Darwin, *Economy of Vegetation*, p. 87 (II, 315–16).

92 Darwin, *Economy of Vegetation*, p. 96 (II, 425–8).

93 Hochschild, p. 243.

94 Darwin, *Economy of Vegetation*, p. 86 (II, 297–300).

95 'Loves of the Triangles', 255–60: see Appendix.

96 Browne, *Darwin: Voyaging*, p. 198.

97 Dick; Priestley, *Sermon*, pp. vii–viii, p. 29.

98 Uglow, *Lunar Men*, pp. 258–62; Inikori, pp. 457–67.

99 Quoted Mason, p. 198.

100 Cowper, pp. 375–6 ('Pity for Poor Africans', lines 5–6, 43–4).

101 Darwin, *Loves of the Plants*, p. 127 (III, 450, 456).

The Temple of Nature

1 My account of *Political Justice* is based on St Clair, pp. 69–84 (quotation p. 78).

2 Locke, quoted Iliffe, p. 431 (from *Essay on Human Understanding*).

3 For relevant accounts of progress, see: Iliffe; Spadafora; Porter, *Enlightenment*, pp. 424–45.

4 Lord Monboddo, quoted Wokler, p. 146.

5 Pope, p. 251 (Epistle II of *Essay on Man*).

6 Quoted Porter, *Enlightenment*, p. 426 (from Gibbon's *Decline and Fall*).

7 Priestley, *Experiments and Observations*, vol. 1, pp. xxxi. See Bindman, *Ape to Apollo*, pp. 58–78, 151–89, and also Wells.

8 Darwin, *Zoonomia*, I, 529.

9 All quotations from the introduction to the first part of 'The Loves of the Triangles': see Appendix.

10 Priestman, 'The Progress of Society?', especially pp. 310–12.

11 Montesquieu, p. 110 (book XV, ch. 5).

12 Augstein; online *Oxford English Dictionary*.

13 Quoted Ellis, p. 107 (from *Introduction to the principles of morals and legislation*, 1789).

14 Jordan, p. 254.

15 Kitson; Wheeler, pp. 235–302; Delbourgo; Ellis.

16 Banton, pp. 13–26; Hudson; Painter, pp. 79–81.

17 Quoted Hochschild, p. 196.

18 Quoted from *Morning Post*, 1785, by Lloyd, pp. 431 and 432.

19 Altink.

20 Quoted Wheeler, p. 4, from *Journal of a Voyage up the Gambia* (1723).

21 Letter to Wedgwood, 22 February 1789, Cambridge University Library, DARWIN 227:1, 110 (reproduced in King-Hele, *Collected Letters*, pp. 331–2).

22 Defoe, p. 145.

23 Defoe, p. 170.

24 Boulukos, especially pp. 1–37 and 75–94.

25 My major sources for this section are Jordan, Bindman, *Ape to Apollo*, and Painter.

26 Acts 17:26 (*King James's version*).

27 Saakwa-Mante, quotation p. 39.

28 Bindman, *Ape to Apollo*, pp. 69–70.

29 Wokler, pp. 152–6.

30 Long, especially vol. 2, pp. 351–404.

31 Pope, p. 247 (*Essay on Man*, I, 207–10).

32 Quoted Jordan, p. 224, from Jenyns, 'Disquisition on the Chain of Universal Being'.

33 William Linn, quoted Jordan, p. 502.

34 King-Hele, *Collected Letters*, p. 181 (to James Watt, 6 January 1781).

35 Delbourgo, p. 2 of typescript (Olaudah Equiano and Thomas Clarkson quoting Dr John Mitchell).

36 Porter, 'Darwin', p. 54.

37 Taylor; Boddice, pp. 25–154 (quote p. 121 from Bentham, *An introduction to the principles of morals and legislation*, [1789]).

38 Pope, p. 197 (*Essay on Man*, I, 245–8).

39 Darwin, *Temple of Nature*, p. 5 (I, 36n).

40 Cloyd.

41 Douthwaite, pp. 11–69.

42 Wokler; Schiebinger, pp. 75–183.

43 Diderot, p. 165.

44 'Loves of the Triangles', footnote to I. 39: see Appendix.

45 Cloyd, p. 162.

46 *Oxford Dictionary of Quotations*.

47 My main sources for this discussion are Jordan, pp. 491–509, and Bindman, *Ape to Apollo*, pp. 201–21.

48 Bindman, *Ape to Apollo*, pp. 201–9 (Camper quoted p. 204).

49 Charles White, quoted Bindman, *Ape to Apollo*, p. 214.

50 Painter, pp. 43–90.

51 Priestman, 'Temples'.

52 Darwin, *Temple of Nature*, pp. 48 and 50 (II, 71–2, 89–90).

53 Priestman, 'Temples'.

54 Darwin, *Temple of Nature*, pp. 33–4 (I, 365–70), and note VI; Priestman, 'Temples'.

55 Darwin, *Temple of Nature*, p. 12 (I, 129–36).

56 Mylonas, pp. 3–22, 224–85. Darwin versifies this myth in *The Economy of Vegetation*, pp. 107–9 (II, 585–610) and pp. 176–7 (IV, 178–94).

57 Darwin, *Temple of Nature*, p. 7 (I, 53–6).

58 Schiebinger, pp. 57–9; Elderkin.

59 Darwin, *Temple of Nature*, p. 3 (I, 1–4).

60 Darwin, *Temple of Nature*, p. 4 (I, 15–18).

61 Darwin, *Temple of Nature*, p. 45 (II, 27–30).

62 Darwin, *Temple of Nature*, p. 61 (II, 243–6).

63 Darwin, *Temple of Nature*, p. 120 (III, 433–4).

64 Darwin, *Temple of Nature*, p. 121 (III, 437–8).

65 Darwin, *Temple of Nature*, p. 134 (IV, 65–6).

66 Quoted Uglow, *Lunar Men*, p. 137.

67 I took both the quotation and the comparison from Dear, p. 74.

68 Vickery, p. 228.

69 Quoted King-Hele, *Darwin: A Life*, p. 89.

70 Secord (Adam Sedgwick quoted p. 371).

71 Darwin, *Zoonomia*, I, 491.

72 Darwin, *Zoonomia*, I, 490. See: Harrison; Wilson, 'Erasmus Darwin'; Zirkle.

73 Darwin, *Zoonomia*, I, 522–3.

74 Quoted Flavell, p. 37 (unpaginated reference to James Hogg's *Confessions of a Justified Sinner*).

75 Roe.

76 Darwin, *Zoonomia*, vol. I, p. 505.

77 Rupke.

78 Darwin, *Temple of Nature*, p. 20 (I, 233–4).

79 Darwin, *Temple of Nature*, p. 26 (I, 295–6).

80 Quoted Porter, *Enlightenment*, pp. 437–8 (Mary Anne Schimmel Penninck).

81 Di Grigorio and Gill, p. 185.

82 Darwin, *Temple of Nature*, p. 54n of the copy CCA.24.64; Darwin, *Loves*, note to I, 65. I owe this discovery to Harman, p. 310.

83 Darwin, *Temple of Nature*, pp. 26–7 (I, 297–302).

84 Darwin, *Zoonomia*, vol. I, p. 505.

85 Note to line 39: see Appendix; Wilberforce, pp. 255–6.

86 Porter, 'Darwin'; Darwin, *Zoonomia*, I, 30–100 (sections IV–XII).

87 Darwin, *Temple of Nature*, p. 28 (I, 309, 313–14).

88 Darwin, *Temple of Nature*, p. 120 (III, 433–4).

89 Letter of 23 January 1816 to Thomas Knight: Banks, *Correspondence*, vol. 6, p. 197.

90 Johnson's *Dictionary*, unpaginated preface (p. 2), citing Richard Hooker.

91 Day, quoted McNeil, *Banner of Science*, p. 71.

92 Letter of 19 January 1790: King-Hele, *Collected Letters*, p. 359.

93 Quoted Cannon, *Life and Mind*, p. 343.

94 Jones, W., p. 69 (from introduction to 'A Hymn to Lacshmi').

95 Jones, W., p. 51 (from introduction to 'A Hymn to Narayena').

96 Jones, W., p. 79 ('The Hymn to Bhavani', from stanza 9).

97 Quoted Cannon, 'Sir William Jones, Sir Joseph Banks', p. 245 (from Jones, 'Anniversary Discourses', 1786). Darwin cited these ideas in *The Temple of Nature*, but he attributed them to Monboddo; Darwin, *Temple of Nature*, p. 5 (I, 36); Cloyd, p. 88, p. 162, n. 13.

98 Knight, pp. 50–1 (III, 37–8, 47–50).

99 Rice-Oxley, p. 64 ('Progress of Man', 72–3, 78–9, 88–9).

100 Holmes, pp. 190–8; Darwin, *Economy of Vegetation*, p. 191 (IV, 373–80).

101 Darwin, Charles, p. 102.

102 King-Hele, *Darwin: A Life*, p. 277.

103 Stevens, p. 116 (22 December 1793).

104 Darwin, *Temple of Nature*, pp. 53–4 (II, 122); Darwin, *Loves of the Plants*, p. 8 (I, 65n).

105 Darwin, *Zoonomia*, I, 509; see Harrison.

106 Darwin, *Zoonomia*, I, 533.

107 Paley, pp. 463–73; *Edinburgh Review* 1 (1802–3), pp. 301–3. See Baldwin, Burbridge, Elliott, Garfinkle.

108 Lewens, pp. 50–5.

109 Quoted Browne, *Darwin*, p. 97 (from Darwin's *Autobiography*).

110 Letter to Watt, 6 January 1781: King-Hele, *Collected Letters*, p. 104.

111 Letter to Watt, 21 June 1796: King-Hele, *Collected Letters*, p. 500.

112 Darwin, *Temple of Nature*, p. 159 (IV, 369–74).

113 McNeil, 'Scientific Muse', pp. 193–7.

114 Malthus, pp. 7–56 (introduction by Anthony Flew).

115 Darwin, *Temple of Nature*, pp. 161 and 163 (IV, 397–400, 425–6).

116 Darwin, *Temple of Nature*, p. 134 (IV, 63–6).

117 Quoted Lewens, p. 10.

118 Montgomery, pp. 39–41.

Conclusion

1 Shelley, p. 432. This edition reproduces the original draft with and without Percy's amendments.

2 Darwin, *Temple of Nature*, p. 7 of 'Additional Note I'.

3 King-Hele, *Darwin: A Life*, p. 369.

4 Porter, 'Darwin', especially pp. 58–60.

5 Browne, *Darwin*, p. 198; Desmond and Moore.

6 Orr, pp. 37–58.

BIBLIOGRAPHY

Note on editions of *The Botanic Garden*

The first edition of *The Loves of the Plants* appeared in 1789, and it was printed in Lichfield. In 1791 the poem was published in London as Part II of *The Botanic Garden*, with *The Economy of Vegetation* as Part I. Darwin altered *The Loves of the Plants* slightly between editions, and the illustrations also changed. The edition I have used is the third, the earliest one in the Cambridge University Library (press mark CCB.47.28). Although the overall title is *The Botanic Garden*, there are two separate volumes: *The Loves of the Plants* (with the original frontispiece by Emma Crewe) and the first edition of *The Economy of Vegetation* (press mark CCB.47.27). At Cambridge, there is another 1791 edition of *The Botanic Garden* (press mark Syn.4.79.13). In that one, both poems are presented in a single volume, and the frontispiece is the second one, by Henry Fuseli.

Abbott, Elizabeth. *Sugar: A Bittersweet History* (London and New York: Duckworth Overlook, 2009)

Aldridge, Alfred O. 'Polygamy and Deism', *Journal of English and Germanic Philology* 48 (1949), pp. 343–60

Alexander, David S. *Richard Newton and English Caricature in the 1790s* (Manchester: Whitworth Art Gallery, 1998)

Altink, Henrice. *Representations of Slave Women in Discourses on Slavery and Abolition, 1780–1838* (London and New York: Routledge, 2007)

Andrew, Donna T. 'Popular Culture and Public Debate: London 1780', *The Historical Journal* 39 (1996), pp. 405–23

Anon. *The golden age, a poetical epistle from Erasmus D—n, M.D. to Thomas Beddoes, M.D.* (London: for F. and C. Rivington; and J. Cooke: Oxford, 1794)

Augstein, H. F. 'From the Land of the Bible to the Caucasus and Beyond: The Shifting Ideas of the Geographical Origin of Humankind', in *Race, Science and Medicine, 1700–1960* (ed. Waltraud Ernst and Bernard Harris) (London and New York: Routledge, 1999), pp. 58–79

Baldwin, John T. 'God and the World: William Paley's Argument from Perfection Tradition—A Continuing Influence', *Harvard Theological Review* 85 (1992), pp. 109–20

Ballantyne, Andrew. *Architecture, Landscape and Liberty: Richard Payne Knight and the Picturesque* (Cambridge: Cambridge University Press, 1997)

Banks, Joseph. *Scientific Correspondence of Sir Joseph Banks, 1765–1820* (ed. Neil Chambers) (6 vols, London: Pickering & Chatto, 2007)

Banton, Michael. *The Idea of Race* (London: Tavistock Publications, 1977)

Baucom, Ian. *Specters of the Atlantic: Finance Capital, Slavery, and the Philosophy of History* (Durham and London: Duke University Press, 2005)

Bewell, Alan. ' "Jacobin Plants": Botany as Social Theory in the 1790s', *Wordsworth Circle* 20 (1989), pp. 132–9

Bewell, Alan. ' "On the Banks of the South Sea": Botany and Sexual Controversy in the Late Eighteenth Century', in David Miller and Peter Reill (eds), *Visions of Empire: Voyages, Botany, and Representations of Nature* (Cambridge University Press, Cambridge, 1996), pp. 173–93

Bindman, David. *The Shadow of the Guillotine: Britain and the French Revolution* (London: British Museum Publications, 1989)

Bindman, David. *From Ape to Apollo: Aesthetics and the Idea of Race in the 18th Century* (London: Reaktion, 2002)

Bindman, David. *Mind-forged Manacles: William Blake and Slavery* (London: Hayward Gallery, 2007)

Binhammer, Katherine. 'The Sex Panic of the 1790s', *Journal of the History of Sexuality* 6 (1996), pp. 409–34

Boddice, Rob. *A History of Attitudes and Behaviours Towards Animals in Eighteenth- and Nineteenth-Century Britain: Anthropocentrism and the Emergence of Animals* (Lewiston: The Edwin Mellen Press, 2008)

Boulukos, George. *The Grateful Slave: The Emergence of Race in Eighteenth-Century British and American Culture* (Cambridge: Cambridge University Press, 2008)

Bradbury, Robert C. *Twentieth-Century United States Miniature Books* (Clarendon, VT: The Microbibliophile, 2000)

Brewer, John. *The Pleasures of the Imagination: English Culture in the Eighteenth Century* (London: HarperCollins, 1997)

Brown, Christopher Leslie. *Moral Capital: Foundations of British Abolitionism* (Chapel Hill: University of North Carolina Press, 2006)

Browne, Janet. 'Botany for Gentlemen: Erasmus Darwin and "The Loves of the Plants" ', *Isis* 80 (1989), pp. 593–61

Browne, Janet. *Charles Darwin: Voyaging* (London: Pimlico, 1995)

Buck, Peter. 'People who Counted: Political Arithmetic in the Eighteenth Century', *Isis* 73 (1982), pp. 28–45

Burbridge, David. 'William Paley Confronts Erasmus Darwin: Natural Theology and Evolutionism in the Eighteenth Century', *Science and Christian Belief* 10 (1998), pp. 49–71

Burke, Edmund. *Reflections on the Revolution in France, and on the Proceedings in Certain Societies in London Relative to that Event* (ed. Conor C. O'Brien) (Harmondsworth: Penguin, 1968)

Cairncross, John. *After Polygamy was Made a Sin: The Social History of Christian Polygamy* (London: Routledge & Kegan Paul, 1974)

Cannon, Garland. 'Sir William Jones, Sir Joseph Banks, and the Royal Society', *Notes and Records of the Royal Society of London* 29 (1975), pp. 205–30

Cannon, Garland. *The Life and Mind of Oriental Jones: Sir William Jones, the Father of Modern Linguistics* (Cambridge: Cambridge University Press, 1990)

Carney, Judith. 'Out of Africa: Colonial Rice History in the Black Atlantic', in Londa Schiebinger and Claudia Swan (eds), *Colonial Botany: Science, Commerce, and Politics in the Early Modern World* (Philadelphia: University of Pennsylvania Press, 2005), pp. 204–20

Carter, Tim. 'Erasmus Darwin, Work and Health', in C. U. M. Smith and Robert Arnott (eds), *The Genius of Erasmus Darwin* (Aldershot: Ashgate, 2005), pp. 289–301

Chamberlain, Robert L. 'George Crabbe and Darwin's Amorous Plants', *Journal of English and Germanic Philosophy* 61 (1962), pp. 833–52

Clarkson, Thomas. *The History of the Rise, Progress, and Accomplishment of the Abolition of the African Slave-Trade by the British Parliament* (1808) (2 vols, London: Frank Cass, 1968 facsimile)

Cloyd, E. L. *James Burnett: Lord Monboddo* (Oxford: Clarendon Press, 1972)

Colley, Linda. *Captives: Britain, Empire and the World 1600–1850* (London: Pimlico, 2002)

Coupland, R. *The British Anti-Slavery Movement* (London: Thornton Butterworth, 1933)

Cowper, William. *The Complete Poetical Works* (ed. H. S. Milford) (London: Oxford University Press, 1907)

Crabbe, George. *The Complete Poetical Works, Volume I* (ed. Norma Dalrymple-Champneys and Arthur Pollard) (Oxford: Clarendon Press, 1988)

Crosland, Maurice. 'The Image of Science as a Threat: Burke versus Priestley and the "Philosophic Revolution"', *The British Journal for the History of Science* 20 (1987), pp. 277–307

Danchin, Pierre. 'Erasmus Darwin's Scientific and Poetic Purpose in The Botanic Garden', in Siergo Rossi (ed.), *Science and Imagination in XVIIIth-century British Culture* (Milan: Edizioni Unicopoli, 1987), pp. 133–50

Darnton, Robert. *The Great Cat Massacre: and Other Episodes in French Cultural History* (London: Allen Lane, 1984)

Darwin, Charles. 'Preliminary notice', in Ernst Krause, *Erasmus Darwin* (London: John Murray, 1879), pp. 1–127

Darwin, Erasmus. *The Botanic Garden; A Poem, in Two Parts* (London: for J. Johnson, 1791)

Darwin, Erasmus. *The Economy of Vegetation (The Botanic Garden, Part I)* (London: for J. Johnson, 1791)

Darwin, Erasmus. *The Loves of the Plants (The Botanic Garden, Part II)* (London: for J. Johnson, 1791)

Darwin, Erasmus. *Zoonomia; or, the laws of organic life* (2 vols, London: J. Johnson, 1794–6)

Darwin, Erasmus. *Phytologia; or the philosophy of agriculture and gardening* (London: for J. Johnson, 1800)

Darwin, Erasmus. *The Temple of Nature; or, The Origin of Society* (London: J. Johnson, 1803) (facsimile edition published by the Erasmus Darwin Foundation, 2003)

Darwin, Erasmus. *To Elizabeth, with Love* (ed. Desmond King-Hele) (Sheffield: Stuart Harris, 2008)

Darwin, Erasmus. *Poems of Lichfield and Derby* (ed. Desmond King-Hele and Stuart Harris) (Sheffield: Stuart Harris, 2011)

Darwin, Erasmus and Anna Seward. *The Pussey Cats' Love Letters: Persian Snow, Po Felina* (Collingswood, NJ: William Lewis Washburn, 1934)

Davies, K. G. *The Royal African Company* (London, New York and Toronto: Longmans, Green & Co., 1957)

Day, Thomas and Bicknell, John. *The Dying Negro, A Poem* (1775) (accessed online)

Dear, Peter. *The Intelligibility of Nature: How Science Makes Sense of the World* (Chicago and London: University of Chicago Press, 2006)

Defoe, Daniel. *The History and Remarkable Life of the Truly Honorable Col. Jacque Commonly called Col. Jack* (London: Oxford University Press, 1965)

Delbourgo, James. 'The Newtonian Slave Body: Racial Enlightenment in the Atlantic World', *Atlantic Studies: Literary, Cultural and Historical Perspectives* 9: 2 (2012), pp. 185–207

De Montluzin, Emily Lorraine. *The Anti-Jacobins 1798–1800: The Early Contributors to the* Anti-Jacobin Review (Basingstoke: Macmillan, 1988)

Desmond, Adrian and James Moore. *Darwin's Sacred Cause: Race, Slavery and the Quest for Human Origins* (London: Allen Lane, 2009)

Dick, Malcolm. 'The Lunar Society and the Anti-Slavery Debate', http://www.search.revolutionaryplayers.org.uk/content/files/18/18/319.txt, accessed 22 October 2007

Diderot, Denis. *Le Rêve d'Alembert* (ed. Paul Vernière) (Paris: Marcel Didier, 1951)

Di Gregorio, Mario, A. and N. W. Gill, *Charles Darwin's Marginalia, Vol. 1* (New York: Garland, 1990)

Douthwaite, Julia V. *The Wild Girl, Natural Man, and the Monster: Dangerous Experiments in the Age of Enlightenment* (Chicago and London: University of Chicago Press, 2002)

Drayton, Richard. 'The Collaboration of Labour: Slaves, Empires and Globalizations in the Atlantic World, c. 1600–1850', in A. G. Hopkins (ed.), *Globalization in World History* (London: Pimlico, 2002), pp. 98–114

Drescher, Seymour. *Abolition: A History of Slavery and Anti-Slavery* (Cambridge: Cambridge University Press, 2009)

Elderkin, George W. 'Diana of the Ephesians', *Art in America* 25 (1937), pp. 54–63

Ellis, Markman. 'Suffering Things: Lapdogs, Slaves, and Counter-Sensibility', in Mark Blackwell (ed.), *The Secret Life of Things: Animals, Objects and It-Narratives in Eighteenth-Century England* (Lewisburg: Bucknell University Press, 2007), pp. 92–113

Elliott, Paul. 'Erasmus Darwin, Herbert Spencer, and the Origins of the Evolutionary Worldview in British Provincial Scientific Culture, 1770–1850', *Isis* 94 (2003), pp. 1–29

Ernst, Waltraud. 'Introduction', in *Race, Science and Medicine, 1700–1960* (ed. Waltraud Ernst and Bernard Harris) (London and New York: Routledge, 1999), pp. 1–28

Fabricant, Carol. 'Binding and Dressing Nature's Loose Tresses: The Ideology of Augustan Landscape Design', *Studies in Eighteenth-Century Culture* 8 (1979), pp. 109–35

Falconer, William. *The Shipwreck: A Poem. In Three Cantos. By a Sailor* (London: for the author, 1762)

Fara, Patricia and David Money. 'Isaac Newton and Augustan Anglo-Latin Poetry', *Studies in the History and Philosophy of Science*, Part A 35 (2004), pp. 549–71

Farrell, Stephen, Melanie Unwin and James Walvin (eds). *The British Slave Trade: Abolition, Parliament and People* (Edinburgh: Edinburgh University Press for the Parliamentary History Yearbook Trust, 2007)

Flavell, Julie. *When London was Capital of America* (New Haven and London: Yale University Press, 2010)

Fox, Celina. *The Arts of Industry in the Age of Enlightenment* (New Haven and London: Yale University Press, 2010)

Franklin, Michael J. *Sir William Jones* (Cardiff: University of Wales Press, 1995)

Fulford, Tim. 'Poetic Flowers/Indian Bowers', in Michael Franklin (ed.), *Romantic Representation of British India* (London: Routledge, 2006), pp. 113–30

Gardner, Bellamy. 'A Chelsea Figure of Aesop', *English Ceramic Circle Transactions* 5/IV (1932), pp. 17–20

Garfinkle, Norton. 'Science and Religion in England, 1790–1800: The Critical Response to the Work of Erasmus Darwin', *Journal of the History of Ideas* 16 (1955), pp. 376–88

George, Sam. *Botany, Sexuality and Women's Writing 1760–1830: From Modest Shoot to Forward Plant* (Manchester: Manchester University Press, 2007)

Godfrey, Richard. *James Gillray: The Art of Caricature* (London: Tate Publishing, 2001)

Good, John Mason. *Lucretius: The nature of things: a didactic poem* (2 vols, London: Longman et al., 1805)

Grafton, Anthony. *The Footnote: A Curious History* (London: Faber & Faber, 2003)

Grenby, M. O. *The Anti-Jacobin Novel: British Conservatism and the French Revolution* (Cambridge: Cambridge University Press, 2001)

Hankins, Thomas L. and Robert J. Silverman. *Instruments and the Imagination* (Princeton: Princeton University Press, 1995)

Harman, P. M. *The Culture of Nature in Britain 1680–1860* (New Haven and London: Yale University Press, 2009)

Harrison, James. 'Erasmus Darwin's View of Evolution', *Journal of the History of Ideas* 32 (1971), pp. 247–64

Harvey, Karen. *Reading Sex in the Eighteenth Century: Bodies and Gender in English Erotic Culture* (Cambridge: Cambridge University Press, 2004)

Hassler, Donald M. *The Comedian as the Letter D: Erasmus Darwin's Comic Materialism* (The Hague: Martinus Nijhoff, 1973)

Herbert, Robert L. *David, Voltaire, 'Brutus', and the French Revolution: An Essay in Art and Politics* (London: Allen Lane, 1972)

Hochschild, Adam. *Bury the Chains: The British Struggle to Abolish Slavery* (London: Pan, 2006)

Hogg, Thomas. J. *The Life of Percy Bysshe Shelley* (2 vols, London: Edward Moxon, 1858)

Holmes, Richard. *The Age of Wonder: How the Romantic Generation Discovered the Beauty and Terror of Science* (London: HarperPress, 2008)

Hopkins, David. 'The English Voices of Lucretius from Lucy Hutchinson to John Mason Good', in Stuart Gillespie and Philip Hardie (eds), *The Cambridge Companion to Lucretius* (Cambridge: Cambridge University Press, 2007), pp. 254–73

Hudson, Nicholas. 'From "Nation to "Race": The Origin of Racial Classification in Eighteenth-Century Thought', *Eighteenth-Century Studies* 29 (1996), pp. 247–64

Hume, David. *Essays: Moral, Political, and Literary* (London: Oxford University Press, 1963)

Hussey, Christopher. *The Picturesque: Studies in a Point of View* (London: Frank Cass, 1967)

Iliffe, Rob. 'The Masculine Birth of Time: Temporal Frameworks of Early Modern Natural Philosophy', *British Journal for the History of Science* 33 (2000), pp. 427–53

Inikori, Joseph E. *Africans and the Industrial Revolution: A Study in International Trade and Economic Development* (Cambridge: Cambridge University Press, 2002)

Jackson, Hazelle. *Shell Houses and Grottoes* (Oxford: Shire Publications, 2002)

Jackson, William. *The Beauties of Nature: Displayed in a Sentimental Ramble through her Luxuriant Fields; with a Retrospective View of Her* (Birmingham: for the author, 1769)

Jay, Martin. *Downcast Eyes: The Denigration of Vision in Twentieth-Century French Thought* (Berkeley and London: University of California Press, 1993)

Jenkins, Ian and Kim Sloan. *Vases & Volcanoes: Sir William Hamilton and his Collection* (London: British Museum Press, 1996)

Johnson, Monte and Catherine Wilson. 'Lucretius and the History of Science', in Stuart Gillespie and Philip Hardie (eds), *The Cambridge Companion to Lucretius* (Cambridge: Cambridge University Press, 2007), pp. 131–48

Johnson, Samuel. *The Rambler* (ed. W. Bate and A. Strauss) (New Haven: Yale University Press, 1969)

Jones, Hazel. *Jane Austen and Marriage* (London: Continuum, 2009)

Jones, William. 'The Enchanted Fruit; or, The Hindu Wife', in Satya S. Pachori (ed.), *Sir William Jones: A Reader* (Delhi, Oxford and New York: Oxford University Press, 1993), pp. 41–8

Jordan, Winthrop D. *White over Black: American Attitudes towards the Negro, 1550–1812* (Williamsburg: University of North Carolina Press, 1968)

Jordanova, Ludmilla J. *Nature Displayed: Gender, Science and Medicine 1760–1820* (London: Longman, 1999)

King-Hele, Desmond. *Erasmus Darwin and the Romantic Poets* (Basingstoke: Macmillan, 1986)

King-Hele, Desmond. *Erasmus Darwin: A Life of Unequalled Achievement* (London: DLM, 1999)

King-Hele, Desmond. *The Collected Letters of Erasmus Darwin* (Cambridge: Cambridge University Press, 2007)

Kitson, Peter. ' "Candid Reflections": The Idea of Race in the Debate over the Slave Trade and Slavery in the Late Eighteenth and Early Nineteenth Century', in Brycchan Carey, Markman Ellis, and Sara Salih (eds), *Discourses of Slavery and Abolition: Britain and its Colonies, 1760–1838* (Basingstoke: Palgrave Macmillan, 2004), pp. 11–25

Klingender, Francis D. *Art and the Industrial Revolution* (London: Palladin, 1968)

Knight, Richard Payne. *The Progress of Civil Society* (London: W. Bulmer, 1796)

Koerner, Lisbet. *Linnaeus: Nature and Nation* (Cambridge, MA and London: Harvard University Press, 1999)

Leask, Nigel. *British Romantic Writers and the East: Anxieties of Empire* (Cambridge: Cambridge University Press, 1992)

Lewens, Tim. *Darwin* (London and New York: Routledge, 2007)

Lloyd, Sarah. 'Amour in the Shrubbery: Reading the Detail of English Adultery Trial Publications of the 1780s', *Eighteenth-Century Studies* 39 (2006), pp. 421–42

Logan, James Venable. *The Poetry and Aesthetics of Erasmus Darwin* (Princeton: Princeton University Press, 1936)

Long, Edward. *The History of Jamaica* (1774) (3 vols, London: Frank Cass & Co., 1970)

Lonsdale, Roger. *Eighteenth-Century Women Poets: An Oxford Anthology* (Oxford and New York: Oxford University Press, 1990)

Madan, Martin. *Thelyphthora* (3 vols, London: J. Dodsley, 1781)

Malthus, Thomas. *An Essay on the Principle of Population* (London: Penguin, 1970)

Marshall, Peter H. *William Godwin* (New Haven and London: Yale University Press, 1984)

Mason, Shena (ed.). *Matthew Boulton: Selling what all the World Desires* (New Haven and London: Yale University Press, 2009)

Mathias, Thomas James. *The Pursuits of Literature: A Satirical Poem in Dialogue* (4 parts, London: J. Owen, 1797)

Matthews, Susan. 'Impurity of Diction: The "Harlots Curse" and Dirty Words', in *Blake and Conflict* (ed. Sarah Haggarty and Jon Mee), pp. 65–83 (Basingstoke: Palgrave Macmillan, 2009)

Matthews, Susan. 'The Surprising Success of Dr. Armstrong: Love and Economy in the Eighteenth Century', in *Liberating Medicine, 1720–1835* (ed. Tristanne Connolly and Steve Clark) (London: Pickering & Chatto, 2009), pp. 193–207

Mavor, [William?]. *The Young Gentleman's and Lady's Magazine* (2 vols, London, 1799–1800)

McKendrick, Neil. 'Josiah Wedgwood and Factory Discipline', *The Historical Journal* 4 (1961), pp. 30–55

McNeil, Maureen. 'The Scientific Muse: The Poetry of Erasmus Darwin', in Ludmilla Jordanova (ed.), *Languages of Nature: Critical Essays on Science and Literature* (London: Free Association Books, 1986), pp. 159–203

McNeil, Maureen. *Under the Banner of Science: Erasmus Darwin and his Age* (Manchester: Manchester University Press, 1987)

Messmann, Frank J. *Richard Payne Knight: The Twilight of Virtuosity* (The Hague and Paris: Mouton, 1974)

Mitter, Partha. *Much Maligned Monsters: A History of European Reactions to Indian Art* (Oxford: Clarendon Press, 1977)

Montagu, Mary Wortley. *The Complete Letters of Lady Mary Wortley Montagu* (ed. Robert Halsband) (3 vols, Oxford: Clarendon Press, 1965–7)

Montesquieu. *The Spirit of Laws* (trans. Thomas Nugent) (Chicago: Encyclopaedia Britannica, 1990)

Montgomery, William. 'Charles Darwin's Thought on Expressive Mechanisms in Evolution', in Gail Zivin (ed.), *The Development of Expressive Behavior: Biology-Environment Interactions* (New York: Academic Press, 1985), pp. 27–50

Montolieu, Maria. *The Enchanted Plants, Fables in Verse* (London: Thomas Bentley, 1780)

Moody, Elizabeth. *Poetic Trifles* (London: T. Cadell & W. Davies, 1798)

Morrell, Jack. 'Professors Robison and Playfair, and the Theophobia Gallica: Natural Philosophy, Religion and Politics in Edinburgh, 1789–1815', *Notes and Records of the Royal Society* 26 (1971), pp. 43–63

Mylonas, George Emmanuel. *Eleusis and the Eleusinian Mysteries* (Princeton: Princeton University Press, 1961)

Nabokov, Vladimir. *Lolita* (Harmondsworth: Penguin, 1955)

Nussbaum, Felicity A. *Torrid Zones: Maternity, Sexuality, and Empire in Eighteenth-Century English Narratives* (Baltimore and London: Johns Hopkins University Press, 1995)

Oldfield, J. R. *Popular Politics and British Anti-Slavery: The Mobilisation of Public Opinion against the Slave Trade 1787–1807* (Manchester: Manchester University Press, 1995)

Orr, David. *Beautiful and Pointless: A Guide to Modern Poetry* (New York: Harper-Collins, 2011)

Ovid. *Metamorphoses* (ed. Mary Innes) (London: Penguin, 1955)

Painter, Nell Irvin. *The History of White People* (New York and London: W. W. Norton, 2010)

Paley, William. *Natural Theology* (London, 1802)

Pearson, Hesketh. *The Swan of Litchfield: Being a Selection from the Correspondence of Anna Seward* (London: Hamish Hamilton, 1936)

Plumptre, James. *The Lakers: A Comic Opera, in Three Acts* (London, 1798)

Polwhele, Richard. *Traditions and Recollections* (London: John Nichols, 1826)

Polwhele, Richard. *The Unsex'd Females: A Poem* (New York and London: Garland, 1974)

Pope, Alexander. *Poetical Works* (ed. Herbert Davis) (Oxford: Oxford University Press, 1966)

Porter, Roy. 'Erasmus Darwin: Doctor of Evolution?', in James R. Moore (ed.), *History, Humanity and Evolution: Essays for John C. Greene* (Cambridge: Cambridge University Press, 1989), pp. 39–69

Porter, Roy. *Enlightenment: Britain and the Creation of the Modern World* (London: Penguin, 2000)

Powell, Nicolas. *Fuseli: The Nightmare* (London: Allen Lane, 1973)

Priestley, Joseph. *A Sermon on the Subject of the Slave Trade; Delivered to a Society of Protestant Dissenters, at the New Meeting, in Birmingham; and Published at their Request* (Birmingham: for the author, 1788)

Priestley, Joseph. *Experiments and Observations on Different Kinds of Air, and Other Branches of Natural Philosophy, Connected with the Subject* (Birmingham: for J. Johnson, 1790)

Priestman, Martin. *Romantic Atheism: Poetry and Freethought, 1780–1830* (Cambridge: Cambridge University Press, 1999)

Priestman, Martin. 'Temples and Mysteries in Romantic Infidel Writing', *Erudit* (February 2002)

Priestman, Martin. 'Introduction', *The Collected Writings of Erasmus Darwin, Vol. I* (Bristol: Thoemmes Continuum, 2004), pp. v–xxvii

Priestman, Martin. 'The Progress of Society? Darwin's Early Drafts for *The Temple of Nature*', in C. U. M. Smith and Robert Arnott (eds), *The Genius of Erasmus Darwin* (Aldershot: Ashgate, 2005), pp. 307–19

Priestman, Martin. 'Lucretius in Romantic and Victorian Britain', in Stuart Gillespie and Philip Hardie (eds), *The Cambridge Companion to Lucretius* (Cambridge: Cambridge University Press, 2007), pp. 289–305

Redford, Bruce. *Dilettanti: The Antic and the Antique in Eighteenth-Century England* (Los Angeles: J. Paul Getty Museum/Getty Research Institute, 2008)

Rice-Oxley, L. *The Anti-Jacobin: or Weekly Examiner. Poetry of the Anti-Jacobin* (Oxford: Basil Blackwell, 1924)

Robinson, Cedric C. 'Capitalism, Slavery and Bourgeois Historiography', *History Workshop Journal* (XX) 1987, pp. 122–40

Robinson, Eric. 'Erasmus Darwin's *Botanic Garden* and Contemporary Opinion', *Annals of Science* 10 (1954), pp. 314–20

Roe, Shirley A. 'Biology, Atheism, and Politics in Eighteenth-Century France', in Denis R. Alexander and Ronald L. Numbers (eds), *Biology and Ideology from Descartes to Dawkins* (Chicago and London: University of Chicago Press, 2010), pp. 36–60

Roper, Derek. *Reviewing Before the Edinburgh, 1788–1802* (Newark: University of Delaware Press, 1978)

Roscoe, William. *The Wrongs of Africa: A Poem* (London: R. Faulder, 1787–8)

Rowden, Frances Arabella. *A Poetical Introduction to the Study of Botany* (London: T. Bensley, 1801)

Rupke, Nicolaas. 'Darwin's Choice', in Denis R. Alexander and Ronald L. Numbers (eds), *Biology and Ideology from Descartes to Dawkins* (Chicago and London: University of Chicago Press, 2010), pp. 139–64

Saakwa-Mante, N. 'Western Medicine and Racial Constitutions: Surgeon John Atkins' Theory of Polygenism and Sleepy Distemper in the 1730s', in *Race, Science and Medicine, 1700–1960* (ed. Waltraud Ernst and Bernard Harris) (London and New York: Routledge, 1999), pp. 29–57

Salinger, Jerome David. *The Catcher in the Rye* (London: Penguin, 2010)

Schiebinger, Londa. *Nature's Body: Gender in the Making of Modern Science* (Boston: Beacon Press, 1993)

Seaton, Beverly. 'Towards a Historical Semiotics of Literary Flower Personification', *Poetics Today* 10 (1989), pp. 679–901

Seaton, Beverly. *The Language of Flowers: A History* (Charlottesville and London: University Press of Virginia, 1995)

Secord, James A. 'Extraordinary Experiment: Electricity and the Creation of Life in Victorian England', in David Gooding, Trevor Pinch, and Simon Schaffer (eds), *The Uses of Experiment: Studies in the Natural Sciences* (Cambridge: Cambridge University Press, 1989), pp. 337–83

Seward, Anna. *Memoirs of the Life of Dr. Darwin: Chiefly during his Residence in Lichfield, with Anecdotes of his Friends, and Criticisms on his Writings* (London: J. Johnson, 1804)

Shelley, Mary. *The Original Frankenstein* (ed. Charles E. Robinson) (Oxford: Bodleian Library, 2008)

Shteir, Ann B. *Cultivating Women, Cultivating Science: Flora's Daughters and Botany in England 1760 to 1860* (Baltimore and London: Johns Hopkins University Press, 1996)

Shteir, Ann B. 'Iconographies of Flora: The Goddess of Flowers in the Cultural History of Botany', in Ann B. Shteir and Bernard Lightman (eds), *Figuring it Out: Science, Gender, and Visual Culture* (Hanover, NH: Dartmouth College Press, 2006), pp. 3–27

Shyllon, Folarin. *Black People in Britain 1555–1833* (London: Oxford University Press, 1977)

Smith, Bernard. *European Vision and the South Pacific* (Oxford: Oxford University Press, 1989)

Snow, Charles. *The Two Cultures* (ed. Stefan Collini) (Cambridge: Cambridge University Press, 1993)

Southam, B. C. *A Student's Guide to the Selected Poems of T. S. Eliot* (London and Boston: Faber & Faber, 1994)

Spadafora, David. *The Idea of Progress in Eighteenth-Century Britain* (New Haven and London: Yale University Press, 1990)

Spary, E. C. 'Political, Natural and Bodily Economies', in N. Jardine, J. A. Secord, and E. C. Spary (eds), *Cultures of Natural History* (Cambridge: Cambridge University Press, 1996) pp. 178–96

Sprat, Thomas. *The History of the Royal-Society of London, For the Improving of Natural Knowledge* (London, 1667)

St Clair, William. *The Godwins and the Shelleys: The Biography of a Family* (London & Boston: Faber & Faber, 1989)

Stevens, William Bagshaw. *The Journal of the Rev. William Bagshaw Stevens* (ed. Georgina Galbraith) (Oxford: Clarendon Press, 1965)

Stones, Graeme. *Parodies of the Romantic Age* (London: Pickering & Chatto, 1999)

Taylor, Thomas. *A Vindication of the Rights of Brutes* (1792) (Gainesville: Scholars' Facsimiles & Reprints, 1966)

Thomas, Hugh. *The Slave Trade: The History of the Atlantic Slave Trade 1440–1870* (London: Phoenix, 2006)

Torrington, Arthur et al. *Equiano: Enslavement, Resistance and Abolition* (Birmingham Museum and Art Gallery, 2007)

Traweek, Sharon. 'Border Crossings: Narrative Strategies in Science Studies and among Physicists in Tsukaba Science City, Japan', in Andy Pickering (ed.), *Science as Practice and Culture* (Chicago and London: Chicago University Press, 1992), pp. 429–65

Uglow, Jenny. *The Lunar Men: The Friends who Made the Future* (London: Faber & Faber, 2002)

Uglow, Jenny. 'But What About the Women? The Lunar Society's Attitude to Women and Science, and to the Education of Girls', in C. U. M. Smith and Robert Arnott (eds), *The Genius of Erasmus Darwin* (Aldershot: Ashgate, 2005), pp. 163–77

Vickery, Amanda. *Behind Closed Doors: At Home in Georgian England* (New Haven and London: Yale University Press, 2009)

Watson, Alexander. *The Anti-Jacobin, a Hudibrastic Poem in Twenty-One Cantos* (Edinburgh, 1794)

Wells, Andrew. 'Race Fixing: Improvement and Race in Eighteenth-Century Britain', *History of European Ideas* 36 (2010), pp. 134–8

Wheeler, Roxann. *The Complexion of Race: Categories of Difference in Eighteenth-Century British Culture* (Philadelphia: University of Pennsylvania Press, 2000)

Wilberforce, Samuel. 'Review of *On the Origin of Species*', *Quarterly Review* 108 (1860), pp. 225–64

Williams, Eric. *Capitalism and Slavery* (New York: Russell & Russell, 1944)

Wilson, Philip. 'Erasmus Darwin on Human Reproductive Generation: Placing Heredity within Historical and *Zoonomian* Contexts', in C. U. M. Smith and

Robert Arnott (eds), *The Genius of Erasmus Darwin* (Aldershot: Ashgate, 2005), pp. 113–32

Wilson, Philip. 'Collecting the Instruments of Life around Me: Anna Seward's Creation of a Life in her Memoirs of Dr Erasmus Darwin (1804)' (Privately published article, 2007)

Wokler, Robert. 'Apes and Races in the Scottish Enlightenment', in Peter Jones (ed.), *Philosophy and Science in the Scottish Enlightenment* (Edinburgh: Donald, 1988), pp. 145–68

Wollstonecraft, Mary. *A Vindication of the Rights of Woman* (1792) (Harmondsworth: Penguin, 1975)

Wood, Marcus. *Blind Memory: Visual Representations of Slavery in England and America 1780–1865* (Manchester: Manchester University Press, 2000)

Wordsworth, William. *The Fenwick Notes of William Wordsworth* (ed. Jared Curtis) (London: Bristol Classical Press, 1993)

Yeazell, Ruth Bernard. *Harems of the Mind: Passages of Western Art and Literature* (New Haven and London: Yale University Press, 2000)

Yohannan, John D. 'The Persian Poetry Fad in England, 1770–1825', *Comparative Literature* 4 (1952), pp. 137–60

Zirkle, Conway. 'The Early History of the Idea of the Inheritance of Acquired Characters and of Pangenesis', *Transactions of the American Philosophical Society* 35 (1946), pp. 91–151

INDEX

Bold entries refer to illustrations.